Phenomenology of the Broken Body

Some fundamental aspects of the lived body only become evident when it breaks down through illness, weakness or pain. From a phenomenological point of view, various breakdowns are worth analyzing for their own sake, and discussing them also opens up overlooked dimensions of our bodily constitution. This book brings together different approaches that shed light on the phenomenology of the lived body—its normality and abnormality, health and sickness, its activity as well as its passivity. The contributors integrate phenomenological insights with discussions about bodily brokenness in philosophy, theology, medical science and literary theory. *Phenomenology of the Broken Body* demonstrates how the broken body sheds fresh light on the nuances of embodied experience in ordinary life and ultimately questions phenomenology's preunderstanding of the body.

Espen Dahl is Professor of Systematic Theology at UiT—The Arctic University of Norway. His research interests mainly focus on the intersection between twentieth-century philosophy (phenomenology and ordinary language philosophy) and theology. His publications include *Stanley Cavell, Religion, and Continental Philosophy* (2014); *In Between. The Holy Beyond Modern Dichotomies* (2011); *The Holy and Phenomenology. Religious Experience after Husserl* (SCM Press 2010). Dahl has published numerous articles on theology and philosophy, such as "Job and the Problem of Physical Pain—a Phenomenological Reading," *Modern Theology* 2016, 32 (1); and "Humility and Generosity: On the Horizontality of the Divine Givenness," *Neue Zeitschrift für systematische Theologie und Religionsphilosophie*, 55 (nr 3) (2013).

Cassandra Falke is a Professor of English Literature at UiT—The Arctic University of Norway. Her books include *Intersections in Christianity and Critical Theory* (ed. 2010), *Literature by the Working Class: English Autobiography, 1820–1848* (2013) and most recently *The Phenomenology of Love and Reading* (2016). She has also authored articles about Wordsworth, Byron, Coleridge, liberal arts education, contemporary phenomenology and the portrayal of violence in literature.

Thor Eirik Eriksen has a PhD in Philosophy and holds a position as senior adviser at The University Hospital of North Norway and Assistant Professor of Community Medicine, at UiT—The Arctic University of Norway. His main research interests are philosophy of science, existential philosophy, phenomenology and the borderland between philosophy and medicine. He has been a contributing author on such articles as: "At the Borders of Medical Reasoning: Aetiological and Ontological Challenges of Medically Unexplained Symptoms" in *Philosophy, Ethics and Humanities in Medicine* (in press), "The Medically Unexplained Revisited" in *Medicine Healthcare and Philosophy* (2012), "Patients' 'Thingification,' Unexplained Symptoms and Response-ability in the Clinical Context" (2016).

Routledge Research in Phenomenology
Edited by Søren Overgaard
University of Copenhagen, Denmark

Komarine Romdenh-Romluc
University of Sheffield, UK

David Cerbone
West Virginia University, USA

Phenomenology of Thinking
Philosophical Investigations into the Character of Cognitive Experiences
Edited by Thiemo Breyer and Christopher Gutland

Wittgenstein and Merleau-Ponty
Edited by Komarine Romdenh-Romluc

Pragmatic Perspectives in Phenomenology
Edited by Ondřej Švec and Jakub Čapek

Phenomenology of Plurality
Hannah Arendt on Political Intersubjectivity
Sophie Loidolt

Phenomenology, Naturalism and Science
A Hybrid and Heretical Proposal
Jack Reynolds

Imagination and Social Perspectives
Approaches from Phenomenology and Psychopathology
Edited by Michela Summa, Thomas Fuchs, and Luca Vanzago

Wittgenstein and Phenomenology
Edited by Oskari Kuusela, Mihai Ometiță, and Timur Uçan

Husserl's Phenomenology of Intersubjectivity
Historical Interpretations and Contemporary Applications
Edited by Frode Kjosavik, Christian Beyer, and Christel Fricke

Phenomenology of the Broken Body
Edited by Espen Dahl, Cassandra Falke, and Thor Eirik Eriksen

For more information about this series, please visit: www.routledge.com/
Routledge-Research-in-Phenomenology/book-series/RRP

Phenomenology of
the Broken Body

Edited by Espen Dahl,
Cassandra Falke, and
Thor Eirik Eriksen

Routledge
Taylor & Francis Group

LONDON AND NEW YORK

First published 2019 by Routledge

2 Park Square, Milton Park, Abingdon, Oxon, OX14 4RN

605 Third Avenue, New York, NY 10017

Routledge is an imprint of the Taylor & Francis Group, an informa business

First issued in paperback 2020

Library of Congress Cataloging-in-Publication Data
A catalog record for this book has been requested

ISBN: 978-1-138-61600-4 (hbk)
ISBN: 978-0-367-73188-5 (pbk)

Typeset in Sabon
by Apex CoVantage, LLC

Contents

Introduction   1

ESPEN DAHL, CASSANDRA FALKE AND THOR EIRIK ERIKSEN

SECTION I
Vulnerable Bodies   11

1   Weakness and Passivity: Phenomenology
of the Body After Paul   13
ESPEN DAHL

2   Perceiving the Vulnerable Body: Merleau-Ponty's
Contribution to Psychoanalyses   29
STÅLE FINKE

3   Torture and Traumatic Dehiscence: Améry and
Fanon on Bodily Vulnerability   49
ALEXANDRA MAGEARU

4   Framing Embodiment in Violent Narratives   66
CASSANDRA FALKE

SECTION II
Suffering Bodies   85

5   Only Vulnerable Creatures Suffer: On Suffering,
Embodiment and Existential Health   87
OLA SIGURDSON

6   The Living Body Beyond Scientific Certainty:
Brokenness, Uncanniness, Affectedness   101
THOR EIRIK ERIKSEN

vi *Contents*

7 No Way Out: A Phenomenology of Pain   119
   CHRISTIAN GRÜNY

8 Toward a Phenomenology of Fatigue   137
   KATHERINE J. MORRIS

SECTION III
Recovery and Life's Margins   155

9 Suffering's Double Disclosure and the Normality
   of Experience   157
   JAMES MCGUIRK

10 Re-possibilizing the World: Recovery from Serious
   Illness, Injury or Impairment   173
   DREW LEDER

11 Notes from a Heart Attack: A Phenomenology
   of an Altered Body   188
   KEVIN AHO

12 Broken Pregnancies: Assisted Reproductive
   Technology and Temporality   202
   TALIA WELSH

13 Dying Bodies and Dead Bodies: A Phenomenological
   Analysis of Dementia, Coma and Brain Death   215
   FREDRIK SVENAEUS

   Bibliography   232
   Contributors   245
   Index   248

Introduction

*Espen Dahl, Cassandra Falke
and Thor Eirik Eriksen*

Phenomenology's account of the body has been one of its major achievements. Phenomenology's primary interest has not been the objective constitution of the body, such as its physical and biological make up, but the body as lived, as enacted and undergone, from the first-person perspective. The phenomenological opening up of new descriptive access to the lived body has proved fruitful both to the phenomenological movement at large and to its many bordering disciplines, such as health sciences, literary studies, theology and aesthetics. In its various attempts to overcome the dualism of mind and body that has haunted Western thinking at least since Descartes, phenomenology has managed to put the mind back into the body and to situate that incarnated subject into its surrounding world. In situating the body in the world, accounts by Husserl, Heidegger and Merleau-Ponty have emphasized active involvement and the capabilities that make such involvement possible. Nevertheless, the lived body also has another dimension, one which shows itself in the impediment of activity, which interferes with our confident relationship to the world and lays bare our dependence on others in it. It is this dimension that this volume calls "the broken body," referring to the body when its activities are interrupted or impeded due to illness, suffering, violence or fatigue. There are many ways in which the lived body can be hindered in its engagement with the world. Brokenness is not limited to a specific class of bodies, but rather occasions disclosure of a vulnerability that belongs to all human bodies. In this book, we propose renewed investigations of the lived body, decidedly starting from the perspective of its broken states.

Choosing "The Phenomenology of the Broken Body" as the title of this volume presumes that three notions will be crucial to its undertaking: phenomenology, the body and brokenness. First, *phenomenology* denotes the philosophical framework of this volume. From this rich and diverse tradition, some concepts have stabilized sufficiently to allow for lines of investigations into, for instance, subjectivity, intentionality, and the life-world. Even as such concepts are fundamental to this book's undertaking, the authors are neither committed to one particular school of phenomenology, nor do they start out from the same disciplinary background.

The collection highlights the importance of inquiry into the broken body for multiple fields and showcases phenomenology's potential for interdisciplinary inquiry. The shared engagement with phenomenology implies certain shared principles, such as the priority of the phenomenal manifestation and its relationship to subjectivity, and certain methods, such as the reduction or reflexive thought along with description. But the exact nature of those principles and methods is not fixed once and for all. The variety represented by the contributors to this book should, we think, be regarded as in tune with the phenomenological spirit. For if the "things themselves," the phenomena as they appear, are the load-stars and yardsticks of phenomenological description, phenomenology must be radically open to them, even as they give themselves differently than prior accounts have had it. Thus, phenomenology requires scholars' willingness to subject ideas to the authority of phenomena themselves, with nothing held absolutely safe except the commitment to a more ade-quate approach. As Maurice Merleau-Ponty has pointed out, perhaps the most important teaching of Husserl is not any one of his principles, but his constant willingness to revise his own thought.[1] Indeed, most of the books Husserl published in his own lifetime were new "introductions" to phenomenology as such, as if his philosophy were never a fixed position but rather a project in the constant movement toward its own realization.

Hence, no school or set of dogmas can contain the phenomenological aspiration; phenomenology, in our view, is a living tradition that takes up its past in constant negotiation with present phenomena. For instance, new insights into the suffering or passive body can make us adjust our understanding of subjectivity and the lifeworld. Moreover, given that phenomena are inexhaustible in their density and saturation, they do not foreclose but rather persistently invoke approaches from various angles and disciplines. In our case, that means that philosophy, theology, health science and literary theory will be presented as participating in the ongoing exploration of the broken body and embodiment as such. The different perspectives presented here demand anchorings in differing concepts of subjectivity, and authors' conceptions of the subject position have affected the style of their phenomenological descriptions as well as the content. For some of the authors, subjectivity is approached from a third-person perspective, while for others, the first-person perspective is adopted as the preferred position from which phenomenological descrip-tion occurs.

Phenomenology, so understood, will be applied to the second central notion of this book, that of the *body*. Practically all accounts of the body, at least within the compass of phenomenology, will today employ the distinction between the outer or objective body (*Körper*) and the expe-rienced and lived body of the subject (*Leib*).[2] As the distinction was ini-tially used by Husserl, the lived body was granted priority because any experience, understanding or knowledge is constituted from the way phenomena are given to someone that is present as an embodied being. For Husserl, this starting point can let us realize how intersubjectivity and the outer world are constituted, or as later phenomenologists have it, how intersubjectivity and the world are already part of the subject from the start. The advantage of the distinction between the objective and lived body is that, without denying the legitimacy of objective sciences, it can preserve the fundamental sense of the embodiment that goes beyond the Cartesian dualism of mind and matter. For the present volume it means that the concept of the lived body facilitates descriptions of fundamental experiences that escape descriptions of the objective body in evidence-based investigations. Pain, for instance, will always appear differently to the sufferer's perspective than any biomedical research can account for. Whatever the cause of broken bodies might be, phenomenology allows for affective impressions, altered world-relation and self-perception to be spelled out, illuminating a too-often overlooked dimension of human experience.

The interest in the body is almost as old as the phenomenological movement, and not by chance, because the subjectivity to which phenomena are given is never ex-carnated, to borrow Chares Taylor's phrase, but embodied—with senses, movements, habits and expressions.[3] Only a few years after what Edmund Husserl called his "breakthrough," in the *Logical Investigations*, he delivered a series of lectures on things and space, devoting much time to analyzing bodily positions and movements that orient us and shape our space. Yet it is in his second volume of *Ideas Pertaining to a Pure Phenomenology* that Husserl's most extensive treatment of the lived body surfaces. Here we find painstaking analysis of the natural substrate of the body, the animated sense of embodiment, with the interplay of different senses, movement and expressivity.[4] However, as both of these works were published posthumously, they naturally did not have any influence in Husserl's lifetime. In spite of the fact that Martin Heidegger, Husserl's one-time assistant, probably knew these works, he did not display any particular interest in the body in his *opus magnum, Being and Time*. However, it has later become clear from lectures to psychiatrists in the 1960s that Heidegger did not neglect the importance of the body—a concern for the body that was already foreshadowed in works prior to *Being and Time*.[5]

The early phenomenological movement in France also developed many congenial insights, as found in Gabriel Marcel and Jean-Paul Sartre. Still, the most significant and standing contribution to the phenomenology of the body is rightly associated with Maurice Merleau-Ponty. Not only did he seek out the mentioned manuscripts in the Husserl Archive prior to publication of his *Phenomenology of Perception*, but he also managed to enrich the phenomenology of the body with the findings of bordering disciplines, especially gestalt psychology. According to Merleau-Ponty, the lived body is not just a vehicle for the ego, but in fact makes up our very being. By means of the body, its sensations, motor intentionality, movements, and habituation, we are never at a distance from the world, but entangled in it from the start. Merleau-Ponty's interest was primarily directed toward the way the active body constitutes our being-in-the-world. Yet he left important suggestions for thinking afresh the meanings and implications of the broken body, drawing on cases from psychology and discussing phantom limbs and compensations for loss of senses.[6] The phenomenology of the body did of course not come to its conclusion with Merleau-Ponty, but has taken on different directions in his aftermath. Michel Henry has explored the immanence of the body and its affective dimensions. Emmanuel Levinas and Jean-Luc Marion draw attention to the body as the site of relationality to others. Lately, even hermeneutics, with its focus on text and interpretation, has taken on a "carnal turn."[7]

As for the third concept central to this book, *brokenness*, as already indicated, refers to a wide variety of experiences, including suffering, pain, passivity, exposure to violence, injury, fatigue and various forms of illness, as well as failing pregnancies and dying or dead bodies. Brokenness suggests a state in which the body is, in passing or permanently, hindered from what was perceived to be its normal functioning, being incapacitated or otherwise debarred from practical or social activities. The first obvious reason for paying heed to the broken state is that a phenomenology of the body should provide thick descriptions of all the modes and inflections that our embodiment involves. In this volume various forms of suffering, illness and marginal states of life are given tentative accounts. Such accounts not only add significantly to the phenomenological discussion, but also provide critical and constructive perspectives on bordering sciences, not least medical sciences.

Apart from providing thick descriptions, attending to the broken body also serves a second aim: it can shed critical light back on what we otherwise take to be the normal body. As Heidegger's famous analysis of the broken hammer has made clear, aspects that tend to be neglected in normal practice, stand out precisely when a tool ceases to function in expected ways. The world suddenly lights up.[8] To be sure, the world is for the most part made accessible to us by means of daily routines and practical engagements along with the orientations which these provide. However, even as they open the world, those daily practices simultaneously narrow it down. Being absorbed in one undertaking, we become blind to the richness and multidimensionality of other aspects that belong to our life in the world. The experience of the body is indeed a case in point. For as we concentrate upon our tasks—riding a bicycle, typing an essay, digging a ditch—the body on which we depend recedes into the background. But when the body breaks down, say by acute pain or a sudden onset of a disease, the background converts into the foreground.[9] Such fundamental interruptions have the power to disclose to us what tends to escape the familiar and habitual body; the breakdown opens the phenomenon up for a new phenomenological access. Importantly, it not only opens up the broken state as such, but also opens otherwise neglected dimensions of our bodily constitution, such as vulnerability, passivity and dependency on the surroundings.

Third, a phenomenology of the broken body enables us to detect the relation between health and sickness, or more generally, between normality and abnormality. That those notions are plastic is not news. Still, some of the subsequent chapters will explore how complex the interweaving between normality and abnormality really is, even how the given body insists on some limits of its plasticity. Not only does the broken body appear as an abnormality, but its brokenness changes what we take as normality, especially as we adapt to new situations. We have a remarkable ability to restore some sense of normality, partly due to biological recovery, but even in lack of full recovery, our ability to adapt to new restrained bodily conditions is striking. New limitations are often transformed into the conditions for new engagements.

The present inquiry into the broken body has in many ways been made possible by earlier work, some of it by scholars we are happy to include in the present volume. In the 1990s, significant contributions to a phenomenology of different abnormal bodily states were made. Taking impulses from the phenomenological movement as a whole and Merleau-Ponty in particular, Kay S. Toombs and Drew Leder paved the way for investigating illness and pain along with other dimensions of the body that had not been sufficiently treated before.[10] Phenomenologists such as Fredrik Svenaeus, Havi Carel and James and Kevin Aho extended these existing routes of inquiry and opened up new ones. These scholars return, not so much to Merleau-Ponty as to Heidegger for inspiration. They also differ from Toombs and Leder in displaying a more direct exchange with bordering medical sciences.[11] The achievements of all these phenomenologists are important backdrops and inspirations for the present book, in which authors consider bodily brokenness with a wider scope, not restricted to health and illness as such. The following chapters work together to clarify relationships between different forms of brokenness, between different representations of brokenness and the experiencing self, and between the self experiencing a broken body and other people. The book's interdisciplinary nature puts it in touch with dimensions central to theology and literary theory as well as philosophy and health sciences.

The present volume is organized into four sections, each containing chapters that draw on different disciplines or angles in order to shed light on the overall theme of the section. In starting with a section called "Vulnerable Bodies," the book addresses some of the underlying conditions of the body often underdeveloped in the classical writings in phenomenology. The vulnerability of the body stems from its exposure to others. While such exposure makes relations and intersubjectivity possible, it also entails the risk of being victim to violence. But the human body is also exposed to itself and its own past, as previous experiences permeate deep structures of the body, for better or for worse. As the vulnerability leads to breakdowns of the body, leaving it weak and deprived of normal functions, it can disclose the otherwise overlooked role of passivity that upholds the body, in weakness as well as in strength.

In the first chapter, Espen Dahl argues that post-Kantian philosophy in general and phenomenology of the body in particular have a tendency to emphasize the active and functional body at the expense of its passivity. Through his reading of Paul's epistles, Dahl finds fruitful insights that point toward the sense and structure of passivity, a sense of passivity that can be further elaborated with the aid of more recent phenomenology. In examining what he calls the bodily unconscious, Ståle Finke shows how past memories and vulnerability to others are central to psychoanalysis as well as phenomenology. Criticizing the Freudian tradition for its lack of attention to the bodily dimensions, Finke employs Merleau-Ponty's phenomenology to enrich psychological accounts. The embodied sense of exposure is pivotal to clinical experience. Therapy deals with recollection, mourning over the past, and acknowledgment of vulnerability to others in order to strengthen individual integrity and separateness. Alexandra Magearu provides an analysis of the effects of colonial violence and individual torture upon the lived body. Not only do to such instances of violence objectivize the victims, depriving them of their voice and reducing them to mere bodies, but they also leave hidden scars within the victims. Such scars will affect their bodily comportment in the world and leave inner wounds that can be re-opened upon further shocks. In this way, the victims of violence display an attitude to the world that goes against the grain of the traditional phenomenology of bodies in the world. Following up this theme, Cassandra Falke brings the role of written narratives into the discussion of potential violence done to the body. The body, Falke argues, is not a tool or transparent medium between self and world, but rather the point of origin for any relation to the world and to other people. From such a phenomenological basis, she examines violent narratives that portray embodiment and relatedness in different ways. The analysis points out how different narratives enliven or deaden readers' awareness of vulnerability and intersubjective connectedness.

The idea of the second section, "Suffering Bodies," is to pay close attention to suffering because suffering is the main way in which brokenness shows itself. Suffering is a broad concept loaded with interpretation and meaning inherited from philosophy, religions and spiritual movements. While it can manifest itself mainly as physical or psychological, suffering is in all cases a total experience. As for the suffering where the physical dimension stands out, it stretches from acute pain, where the sense of vitality can become unbearable, to fatigue, where one typically feels drained of all vitality. Neither pain nor fatigue simply happens to the objective body; both reveal themselves from within the experiencing subject, who finds the embodied present and the possibility for future experiences radically realigned by an altered bodily state.

In addressing the relation between suffering, embodiment and health, Ola Sigurdson argues that suffering should be understood as a mode of being in the world. Suffering, the chapter argues, is essential for understanding human beings as vulnerable creatures. It is through suffering as an active passivity that the experience of pain or other modes of suffering can be transformed to a constructive relationship that could be called existential health. Thor Eirik Eriksen also holds a wide understanding of suffering and brokenness and pays particular attention to the affectedness and uncanniness that tend to accompany such brokenness. In so doing, Eriksen argues that the notion of objectivity and object-ness that prevails in much current science, fails to account for how brokenness, uncanniness and affectivity belong to the human condition. Christian Grüny focuses on one central manifestation of suffering, namely physical pain. Pain, he argues, is a complex experience that should be understood as a process within our interaction with the world, a specific mode of sensing rather than a single quality or entity. The phenomenological description Grüny works out calls into question the idea of a unitary lived body and reveals it as heterogeneous. In her chapter, Katherine J. Morris investigates fatigue, which, unlike pain, has hardly been given any attention within phenomenology. Morrison provides an analysis of tiredness and fatigue in their ordinary manifestations, alongside an account of chronic fatigue syndrome, with its similarities and differences from ordinary fatigue. On the basis of her analysis of Chronic Fatigue Syndrome, Morrison invites a dialogue between phenomenology and medical sciences, and argues that phenomenology will profit from such dialogue.

Section three, "Recovery and Life's Margins" corrects the impression that the study of the broken body concerns itself exclusively with breakdowns, inhibitions, suffering and passivity. Restitution, recovery and adaptation also play important roles. This section sheds light on the dynamics that unfold between the conceptualization of abnormal versus normal, breakdown versus recovery. In bringing up the relation between such concepts, it is our aim to account for how surprisingly complex this relation is: bodily breakdowns can disclose hitherto hidden dimensions of the normal lived body. But the abnormality of illness can also call forth resources from the bodily repertoire to create a new normality. Indeed, the capacity to recover from the onset of a radical disruptive illness and constantly adapt indicates a remarkable power of the lived body. As phenomenology's and existentialism's interests in limit phenomena often implies, the encounter with margins has power to shed light on the very center of our being in general, and indeed our embodied being. The margins investigated in this section include not only normality and abnormality but also the margins of life, death, near-death and brain death.

James McGuirk looks into how different strands of phenomenology have dealt with illness and suffering as modes of abnormality. For all the powers to regain wholeness and integrity, normality and abnormality are not open to endless constructions; for, as McGuirk argues, the body sets limits for what can be perceived as normal and the loss of normality. Drew Leder examines the resources we have for adaption and restoration. Even if the broken body can appear as an inhibition, something that blocks us from pursuing our goals, there are ways in which we tend to cope with the new situation. Recovery may not involve a medical cure and may take place in the face of ongoing impairment. Such recovery is accounted for with respect to embodied and existential resources as well as spiritual resources for self-transcendence. Kevin Aho draws on his first-person experience of undergoing a heart attack to illuminate the marginal experience of being torn out of ordinary being-in-the-world. In describing how serious illness and proximity to death affect structures of spatiality, temporality, understanding and intersubjectivity, Aho points out how illness can also initiate our capacities for restoration and healing.

While birth and death are universally shared by humans, they nevertheless raise culturally specific questions. In particular, as the medical and technical possibilities in our culture force ethical and political problems to the fore, these problems in turn refer us back to the basic question of what a living body is. Talia Welsh provides reflections on pregnancy, or more precisely, the absence thereof in the time of assisted reproductive technologies. Such technologies provide a new horizon regarding pregnancy and make infertile bodies appear not as broken, but rather as sites of a temporary problem that can be solved with the right medical intervention. Such a technological horizon has wide-ranging consequences, not only for our view of pregnancy, but also for the perception of the broken body: what was previously regarded as a destiny now becomes a curable problem. Turning from pregnancy and birth to death, Fredrik Svenaeus begins his chapter by asking whether a brain dead person can be declared dead. The issue concerns ethical aspects of artificial prolongation of life and the possibility of organ transplantation. More central to phenomenology, Svenaeus demonstrates how it also concerns our understanding of the relationship between a human body and personhood. He asks if the lack of activity in the brain as a definition for death is at odds with a phenomenological understanding of personhood. Such definition can, moreover, be taken as an expression of an instrumentalization that leads to the perception of the human body as a container of organs.

This collection of analyses, descriptions, examinations and arguments concerning the broken body sheds further light on a topic relevant for various disciplines. It draws on the reservoirs of the phenomenological tradition and attempts to expand phenomenological inquiry toward a richer, more self-reflective and critical approach to the living bodies that we are.

Notes

1. Maurice Merleau-Ponty, *Themes from the Lectures at the Collége de France 1952–1960*, trans. John O'Neill (Evanston, IL: Northwestern University Press, 1970), 105–8.
2. Edmund Husserl, *The Crisis of European Sciences and Transcendental Phenomenology*, trans. David Carr (Evanston, IL: Northwestern University Press, 1970), 107.
3. Charles Taylor, *A Secular Age* (Cambridge, MA: The Belknap Press of Harvard University Press, 2007), 615, 741.
4. Edmund Husserl, *Ideas Pertaining to a Pure Phenomenology and to a Phenomenological Philosophy*, Second Book, trans. Richard Rojcewicz and André Schuwer (Dordrecht: Kluwer Academic Publishers, 1989), Section Two.
5. Martin Heidegger, *Zollikoner Seminare*, ed. M. Boss (Frankfurt am Main: Vittorio Klostermann, 1987), 97–120; *Basic Concepts of Aristotelian Philosophy*, trans. R.D. Metcalf and M.B. Tanzer (Bloomington, IN: Indiana University Press), 129–40. For elaboration of these themes, see Kevin Aho, *Heidegger's Neglect of the Body* (New York: State University of New York Press, 2007).
6. Maurice Merleau-Ponty, *Phenomenology of Perception*, trans. Colin Smith (London: Routledge, 1992), Part One.
7. Michel Henry, *Incarnation: A Philosophy of the Flesh*, trans. Karl Hefty (Evanston, IL: Northwestern University Press, 2015). Richard Kearney and Brian Treanor, eds., *Carnal Hermeneutics* (New York: Fordham University Press, 2015).
8. Martin Heidegger, *Being and Time*, trans. Joan Stambaugh (Albany: State University of New York Press, 2010), 72–6.
9. Drew Leder, *The Absent Body* (Chicago: The University of Chicago Press, 1990), 71.
10. Ibid.; Kay S. Toombs, *The Meaning of Illness: A Phenomenological Account of the Different Perspectives of Physician and Patient* (Dordrecht: Kluwer Academic Publishers, 1992).
11. Fredrik Svenaeus, *The Hermeneutics of Medicine and the Phenomenology of Health: Steps Towards a Philosophy of Medical Practice* (Dordrecht: Kluwer Academic Publishers, 2000); Havi Carell, *Illness: The Cry of the Flesh* (London: Routledge, 2014); James Aho and Kevin Aho, *Body Matters: A Phenomenology of Sickness, Disease, and Illness* (Lanham: Lexington Books, 2008).

Section I
Vulnerable Bodies

1 Weakness and Passivity

Phenomenology of the Body After Paul

Espen Dahl

How is the body given? This is a fundamental question for phenomenology since givenness is nothing but the way in which something is offered up for intuition, in other words, the phenomena, the "things themselves." If we are to investigate the body phenomenologically, there is nothing else to appeal to, for the originary givenness—givenness in the flesh (*leiblich gegeben*) as Husserl says—is the very source but also the limit of phenomenological investigation.[1] According to our everyday, natural attitude, we speak of bodies as material entities available from a third-person perspective, in German: *Körper*. There is, however, another way of givenness that is crucial to phenomenology, the body as felt, exercised and suffered from the first-person perspective: *Leib*. The importance and fruitfulness of this latter notion cannot be underestimated; it has opened the path for research into the lived body that would otherwise remain outside philosophical conceptions.

Yet, as I will try to show, the traditional phenomenological account of the lived body takes for granted a capable and achieving subjectivity in a way that implies, I believe, an activist bias that has some serious consequences for our understanding of the role of activity of body, the valuation of passivity and the self. I want to propose that there is also another sense of bodily givenness, one which is not dependent on activity, but which takes the strict meaning of "givenness" as something to be received passively. As a guide to the phenomenology of the broken body, I will draw on Paul, who, in my reading, points toward a profound sense of receptive passivity. Such perspective, I will finally argue, points toward an important sense in which embodiment and subjectivity can be conceived.

I Can

Since phenomenology must submit itself to the way things themselves are given, and hence turn to the lived body, it is not by accident that it has tended to start with the way the body usually gives itself in perception, in movement and in action. Let me, with broad strokes, point to the importance of the active, achieving subjectivity, as found in Husserl, Merleau-Ponty and (more implicitly) in Heidegger.

One of the epoch-making contributions both put forward by Husserl and Merleau-Ponty is comprised in the little formula: "I can."[2] It alludes to the basic tenets of modernity, first put forward by Descartes in his "I think," which is the indubitable foundation of self and world, and later elaborated in Kant's "I think," the transcendental apperception that assures the possibility of the unity of experience, anchored in subjectivity. The problem with those prior conceptions is that they tend to split consciousness from the world. In order to bridge the ontological gap, they turn the world into representation, a representation that is constituted by means of sensational intake and intellectual synthesis and interpretations. This, of course, fits into the way scientists investigate the empirical world by construing general theories and isolated objects and stepwise testing the theories against empirical facts. However, such a maneuver deforms both the lived world-relation and the role of the constituting subject that is engaged in that world.[3] In short, it hardly accounts for our being-in-the-world.

In turning "I think" into "I can," Husserl and not least Merleau-Ponty are making two fundamental shifts. First, they turn the question of subjectivity and world away from the intellectual conception: "It is, rather, a kind of practical synthesis: I can."[4] The surrounding world is not brute nature, but a perceptual field of forces that appeals to us and to which we respond.[5] We do not so much observe the world as we pursue diverse tasks in it. Each new situation calls upon new acts of perception, acquired skills and prior experience in order to respond to the question that arises therein. Second, "I can" is neither the isolated ego cogito nor a transcendental apperception, whether that is conceived as a soul, mind or otherwise, but already an embodied subjectivity. The body is not something I have or inspect, but the place where the I is originally at home—I am my body. That the body as such opens the world, is only possible because the body is not conceived of as a thing, but as always already ensouled, or with Husserl: "Man, in his movements, in his action, in his speaking and writing, etc., is not a mere connection or linking up of one thing, called a soul, with another thing, the Body. The Body is, as Body, filled with the soul through and through."[6]

The importance of the body in the recasting of philosophical problems is due to its role in overcoming the enigmatic relation between consciousness and an external world. As consciousness is a way of being in the body, the world never appears foreign: for the body irreducibly belongs to both consciousness and world, as a third dimension. It is bestowed with different senses: with sight in which colors, forms and distance are given; with the ability to hear sounds with different qualities, pitches and volumes; and with tactility that makes us able to touch and feel. We can reach out and touch tables, cups, handles and even ourselves. Indeed, touch plays a remarkable role in the self-awareness of the lived body because touch not only registers the shape and quality of the objects, but at the same time, it leaves impressions that can be felt on the body that touches. Any touch is thus "double sensation"—which is, moreover, doubled up when the right hand touches the left.[7] Unlike seeing and hearing, touch's double reference awakens the body to its immediate self-awareness.

The lived body can also move. The free ability to realize potentiality, the "I can," is intimately interwoven with motility. When I reach out for a coffee cup, I do not measure out the distance and direction in geometrical space, but direct myself intentionally toward the desired cup—as a task that I accomplish by means of my motor intentionality. Part of what makes this possible is without a doubt my acquired habits, which bring my past experiences into the present project directed toward a future goal. But for any meaningful movement to take place, such movement must also be controlled from this side of the lived body, and hence we are equipped with a particular sensation that always accompanies movements, so-called kinesthesis.[8] That there is an internal relation between movement, kinesthesis and other sensations is obvious from the simple fact that we constantly move our eyes, head, hands and feet as we perceive. Notably, it is not by means of intellectual synthesis that those sensations and movements are fitted together as if constructing the object we perceive. A unity is entailed in the overarching animation of the entire moving body, that is, by a bodily synthesis in a constant movement toward the world.[9] Such a fundamental body-intentionality is both intersensoric and practical at the same time. As the vehicle or, as Husserl prefers, organon of my subjectivity, the body is expressive of the "I can" implied in all intentionalities. This holds not only for perception but also for practical life—the sphere of freedom, will and actions. "I can" is the very original capability ("can") that animates the body, not as causality but as expressive of my motivations ("I"). Husserl writes:

> With regard to my central Ego-acts, I have consciousness of the *I can*. They are indeed activities, and in their entire course we have precisely not a mere lapse of events but, instead, the course they take continuously proceeds out from the Ego as center, and as long as that is the case, there reigns the consciousness, "I do," "I act."[10]

This underlines the capabilities, activities and achievements that the "I can" conducts in exploring my projects and perceptions of the world, comprising all the potentialities and the freedom of will that are at my disposal. "I can" unites subjectivity and body, but also gives it an activist orientation: it is by realizing my potentiality that I accomplish the body that I am.

The centrality of ability and activity is perhaps even more salient in Heidegger's analysis of Dasein, which takes on a more existential dimension.

Being-able-to-be (*Sein-können*) belongs, according to Heidegger, to the Being of Dasein, as one of its founding existential structures (*Existenziale*). It this way, Heidegger makes Dasein's Being inextricably tied to its active abilities. It comes to the fore in various dimensions of our existence, such as in our ability to understand, as we always understand in light of a future goal or projection. We are indeed thrown into the world—the sheer facticity of our existence—but we have a horizon of future possibilities that opens up a room for what we are able to achieve.[11] The active Being-able-to-be thus aims toward the future, it is forward looking in the sense that it points to who we can possibly become, which in turn gives meaning to who we presently are and have become. However, Dasein tends to fall into the inauthentic mode of being in which its innermost Being-able-to is leveled out and made to conform to "the they" in the anonymous everyday existence. But in anguish, the call of the conscience, and most clearly, in Heidegger's analysis of Dasein's relation to his/her own death, the true possibilities are opened for our understanding, and Dasein's innermost Being-able-to is uncovered as a possibility for authentic existence. Being-toward-death thus tears Dasein away from its average existence of the They; it individualizes it, as death is always inescapably mine.[12] The anxiety that death evokes has the paradox of impossibility and possibility and yet empowers the Being-able-to-be: "In anxiety, Dasein finds itself *faced* with the nothingness of the potentiality-of-being [Being-able-to-be/*Seinkönnen*] of the being thus determined, and thus discloses the most extreme possibility."[13] Death, by means of its impossibility, becomes the impetus of renewed abilities to be.

I take such insights to be major contributions to the phenomenology of the body and to the understanding of the active embodied person. But it comes with a price. For, despite all the differences between Husserl, Merleau-Ponty and Heidegger, they share an underlying activist bias, where the identity of the embodied I, together with the shaping of his/her lifeworld, seems to presuppose the active, well-functioning body. It is worth asking whether such phenomenological activism eclipses another dimension of the body, namely the passive dimension. The problem first arises when the embodied person for some reason breaks down, is not any longer able to act freely or fulfill his/her potentiality or is otherwise not functioning in the normal sense.

Nevertheless, there seems to be a sense in which Husserl, Merleau-Ponty and Heidegger are highly sensitive to passivity. Husserl elaborates with painstakingly patience the way the active comprehension only comes to pass provided that a passive stratum of experience is pre-given; not only pre-given, but it somehow awakens the activity in the first place, by its allures or appeals, to which the active apperception is a response. At times he can speak of how the active pole of the ego is confronted with an affectivity that refers all the way back to the givenness of the original impressions.[14] However interesting this line of thought is, it still trades on the basic pattern of an active pole that certainly does not create its matter, but still forms the passively given matter in a process he calls constitution. Something similar, I take it, is true of Merleau-Ponty, for indeed neither perceptual objects nor the ego is transparent on his account, but inexhaustible in depth, thickness and anonymous meaning that stem from passive strata of experience. To recognize passivity, manifest in sleep, in our past or in the givenness of the world, Merleau-Ponty writes, means to realize "a field of existence already instituted, which is always behind us and whose weight . . . only intervenes in the actions by which we transform it."[15] However, since the passivity is transformed through actions, passivity always refers back to the activity of "I can." As for Heidegger, despite the emphasis being put on Being-able-to-be, it is not a matter of being able to do anything without constraints—that would render Heidegger's notion of freedom completely vacuous. First of all, Dasein is thrown into existence. This facticity means that Dasein is always already exposed to its existence prior to any activity on its part. Even if facticity primarily means the fact that you are thrown into existence, Heidegger is careful to point out that he is not talking about isolated existence, but rather an existence always passively given along with and restrained by the world and others in it.[16] Facticity makes up the given condition from which the projections of Dasein's own possibilities can be sketched out and pursued, and in this sense only underpins the importance of the abilities and activities of Dasein. Even death, the very limit of all activity, becomes, in Heidegger's analysis, the "most extreme possibility."

In sum, the sense of passivity that is operative in traditional phenomenology is conceived as the dialectical counterpart of activity, which provides certain raw material or resistance without which the phenomenological accounts would fall back on idealism. The problem is that this hardly accounts for the inability—"I cannot"—that is central to the phenomenology of weakness and brokenness. Havi Carel has made the convincing argument that, despite the fruitfulness of Heidegger's Being-able-to-be, it is futile when it comes to accounting for impaired life. If we want to follow Heidegger, Carel suggests, his phenomenology must be supplemented on this point, so that "inability to be" can be recognized as more than a deficient mode of Being-able-to-be. "Acknowledging an inability and learning to see it as part of life's terrain are important lessons that illness can teach at any age. This knowledge enables the ill person to embrace the unable self as part and parcel of human existence."[17] But can phenomenology achieve this as long as it takes the "I can" as its given starting point?

Passivity and Breakdowns

There might, however, be another dimension of passivity, one that does not so much concern that passivity dialectically corresponding to practical or epistemic activities, but which nevertheless belongs to the body itself. If this is so, then the very activist assumption, namely too-tight identification of embodied subjectivity and activity, must be called into question. The passive dimension I have in mind shows up in abnormality, illness, breakdowns—in short, in the various ways in which we are exposed to "I cannot." Such breakdowns presuppose a normal functionality being brought to a halt.

The reason why, for instance, Heidegger invests interests in breakdowns is not so much because it opens up a new phenomenological field of the body, but rather its power to uncover dimensions that otherwise go unnoticed. Given the primacy of the active engagement in the world, Heidegger claims, we tend to be absorbed by the very aim of our directedness. That means that those things which are most near, such as our tools, tend to withdraw from thematic sight. It is, according to Heidegger's analysis, when the tools break, show themselves as unusable or even obstructive to our dealings that they first "light up." We see the tools we took for granted laying there; we understand the entire network of assignment relations that were operative. The interruptions, failures and breaks make the previously unthematic world announce itself.[18] Heidegger's phenomenological analysis of tools has also applicability for the way the body shows itself, since the body also breaks down, in pain, fatigue, aging and illness. As long as the lived body functions in the normal way, it does not draw attention to itself, but rather leads away from itself, as if it is the silent background from which a foreground gets into focus. Heidegger picks this up in a much later seminar, where he says that the body is an enigmatic phenomenon since it is away from itself as it gets absorbed in its tasks.[19] This absent body, to borrow Drew Leder's terminology, is precisely what is brought to presence in breakdowns: it changes from background to foreground, from disappearing to intruding presence or "dys-appearing."[20]

If we turn to Merleau-Ponty, it is also true that he explores different forms of disease and illness. For instance, there is a lengthy discussion of phantom limbs: how can it be that someone can still feel a limb when it is no longer there? There is, according to Merleau-Ponty, no satisfactory mechanical or psychological explanation for it; nevertheless, there is a phenomenological one. The habitual body, structured by the body sedimentations of past acquirements, continues to direct itself toward the world and its tasks, even if such directions no longer correspond to the actual body. The phantom limb depends on a kind of repression of the actual body and actual situation. The upshot of this is, by means of its contrast, that we normally make the world ours, not by psychic or physiologic means, but by drawing on habitualities that adjust us to various circumstances.[21]

On another occasion, Merleau-Ponty draws on a case from psychiatric literature, namely the case of Schneider, who had suffered a physical damage in the back of his head and subsequently suffered from strange disorders. Schneider is able to perform concrete operations and concrete demands, but has lost the ability to perform abstract movements (such as drawing a circle in the air without watching his own limbs) or imagine himself in non-actual situations. Moreover, Schneider understands messages and movements, but is unable to see their immediate implications. His movements have lost the original inherent motor intentionality by means of which we perceive and handle the more abstract possibilities that the world opens to us. Schneider has lost the way in which movements and perception together draw on the past as a background from which we can project ourselves into new, open situations. In other words, he has lost what Merleau-Ponty calls the "intentional arc."[22] To him, the world appears frozen and readymade, his actions remain fixed to the present and are performed in an almost mechanical way. The point Merleau-Ponty is making is that Schneider's way of compensating for this loss sheds light on the roles motor intentionality and the intentional arc normally have. Merleau-Ponty nevertheless does warn us not to draw any hasty conclusions from the abnormal to the normal bodily existence, since the former makes up a world on its own. Keeping that in mind, there is nevertheless an indirect way in which pathological cases shed light on the normal functioning "I can"—which is, after all, Merleau-Ponty's overarching purpose: "We must take substitutions as substitutions, as allusions to some fundamental function that they are striving to make good, and the direct image of which they fail to furnish."[23]

Perhaps Husserl should not be brushed aside too quickly either as a phenomenological activist; it is indeed Husserl who first brings up the notion "I cannot." Discussing the matter, Husserl initially points out that there are different grades of "I cannot." There are external occasions where "I can" meets different degrees of external resistance that can be overcome with lesser or greater efforts, but there are also occasions where the resistance is insurmountable. In both cases, the resistance at stake is not meant as materiality that casually opposes my objective body, but is experienced within the lived body as oppositions to my will.[24] More interesting in my respect, however, are cases where the "I cannot" is due to inabilities of the lived body itself. Husserl again points out that there are different grades of "I cannot" by means of a series of examples: I used to play the piano, but I have not practiced for a long time, and the piece is no longer within reach. In this case, "I cannot" can be overcome with some practice. However, with other practices, such as typewriting or cycling, lack of constant practice does not lead to a loss of mastery; I can usually pick up writing or cycling again and display the same degree of mastery.

From Heidegger's, Merleau-Ponty's and Husserl's use of broken bodies, we can clearly see that they in no way try to shy away from the vulnerability of human capabilities. Yet despite their willingness to pay attention to bodily breakdowns, they do not disclose a willingness to dig into the passivity that the broken body exposes. Rather they all seem to be dominated by a particular methodical aim: the broken body sheds light back on the otherwise inaccessible dimension of the normal functioning body. With regard to the dichotomy of normality and abnormality implied in health and illness, Husserl says that "this modified givenness refers back to the normal."[25] "I cannot" leads phenomenology invariably back to "I can." Another way of putting this is to say, with Carel, that illness works as a kind of phenomenological *epoché*. The point is still the same: by disrupting our ordinary activity, illness makes what is otherwise implicit, explicit and thematic. Husserl's notion of *epoché* means bracketing the natural attitude in which we take ourselves and our world for granted, in order to turn our reflective attention to the phenomena, their structure, mode of givenness and constitution. In Carel's words: "The epoché asks us to dislodge ourselves from everyday habits and routines in order to reflect on them; this, I suggest, is what happens in illness, albeit in a raw and unformulated manner."[26] In this way, we can say that Merleau-Ponty's pathological cases, Husserl's "I cannot" and Heidegger's breakdowns all functions as occasions for *epoché*.

However, the *epoché* thus conceived also retains a less obvious but still remarkable bias toward activism. It leads back to the normal, well-functioning body, indeed, toward "I can" from which it departs. One might suspect that a conception of the embodied subjectivity that is modeled on the well-functioning, achieving person lacks the resources for integrating the weakness and brokenness as such--their givenness in their own phenomenological terms. Tellingly, when Husserl considers the case of severe nerve disorder, in which "I cannot" is made complete, he does not offer a phenomenology of the passive body, say as integral to our embodied condition, but pushes it outside the established sense of bodily subjectivity: "In that respect I have become an other."[27] Such statement is quite consistent with his emphasis on "I can"—where ego, body, perception and activity make up one identity, the I of "I cannot" has no place; it must be split off, becoming another I.

The question is whether it is possible to follow the lead of *epoché* in the opposite direction, so to speak, not from breakdowns back to the normality of "I can," but deeper into the structures and givenness of "I cannot" itself. If it is possible to lay bare the passivity itself exposed in the broken body, it is furthermore important to ask whether such utter passivity is not a stranger to the self, but makes up one of its fundamental structures, from which activity in turn flows. Such an understanding would need a sense of passivity that is not only conceived as dialectical correspondent to activity, but rather some deeper form of "I cannot" that is found beneath that dialectic. I think the letters of Paul contain important insights in this respect.

Paul on Strength and Weakness

There is no doubt that the body is central to Paul, both as the bearer
of the self, a metaphor for the congregation, and as the whole human
person in activity and passivity. Indeed, the body makes up a point from
which all the major topics of Paul's anthropology can be oriented: the
law and desire, members and unity, sin and redemption, sin and holiness,
mortality and immortality.[28] However, Paul's letters lend themselves to a
phenomenological reading that corresponds neatly to the topic under discussion, such as the capable, active and strong body versus the incapable,
passive and weak body, particularly as it turns up in the context of The
Second Letter to the Corinthians.

In the relevant passages, the context is made up by Paul defending his
authority as an apostle over against rivaling missionaries that must have
challenged his status, obviously boasting of their spiritual, social and
bodily supremacy. Hence, Paul defends his legitimacy and also deals with
the charges of being weak while present in Corinth, whatever that might
refer to (2. Cor. 10: 10). Paul's strategy is surprising: instead of defending himself from the charge of being weak, he in fact comes to insist on
it. He turns around the criteria of evaluation, reversing the established
hierarchy that is taken for granted, that is, strength over weakness, activity over passivity:[29] "If I must boast, I will boast of the things that show
my weakness." (2. Cor. 11: 30). The rhetorical move is clear; apparently
competing with his opponents in boasting, Paul in effect deconstructs
all boasting of strength. Yet he can still claim that the only legitimate
boasting is "boasting in the Lord" (10:17), but of course, boasting is per
definition self-referential—not referring to the reception of a message
about someone else, which Paul appeals to. This double reversal—from
strength to weakness, self-reference to reference to Christ—takes its
final legitimation from the cross. If the cross is the criterion of apostleship, then any legitimate apostle must show himself, not in strength, but
precisely in weakness—and paradoxically, that will be Paul's strength.[30]
Hence in Paul's mad boasting, there is thus both a positive reference to
the strength of the Lord, and a negative, in boasting of his weakness.
How do these fit together?

This paradoxical structure is pushed further in Chapter 12. For here
too there is another paradoxical act of boasting, propelled, one must
assume, by Paul's opponents' claim to spiritual superiority and strength.
As if competing with his opponents, he testifies to his ecstatic vision,
which only a few in the Hebrew Bible have been granted. He has once
been, as he writes, "caught up to the third heaven—whether in the body
or out of the body I do not know; God knows. And I know that such
a person—whether in the body or out of the body I do not know; God
knows—was caught up into Paradise and heard things that are not to
be told, that no mortal is permitted to repeat" (2. Cor. 12: 2–4). The

intention of accounting for this vision is, however, precisely not to hammer home his spiritual strength or superiority, since any authority of his cannot, according to his line of argument, be based on such boasting. Even if such mad boasting makes Paul the equal, if not superior of his opponents, the point is that, as a criterion of apostleship, the ascent misses the mark. So the emphasis is rather on his descent. Indeed, to prevent the ascension from making the ego inflated, Paul is violently brought down to earth. "Therefore, to keep me from being too elated, a thorn was given to me in the flesh, a messenger of Satan to torment me, to keep me from being too elated" (12: 7). It is as if the thorn pricks the inflated self, like a puffed up balloon, and returns him, deflated, firmly into his own body.[31]

The thorn in the flesh is one of the most famous expressions in Paul's epistles, and much ink has been spilt on pinning down what exactly the thorn refers to. Already from the early church onwards, there have been different theories put forward, such as some form of physical illness—epilepsy, eye problems, stammering or malaria—to some sort spiritual trial, or even persecution during his missionary work.[32] The exact reference is, however, not decisive; I am just relying on the most common and convincing reading that takes it to be connected directly to his flesh, as some form of affliction, disabling disease, that is, a bodily experience of "I cannot."[33] It is for this reason that the repeated appeal to weakness now takes on a very specific carnal meaning, as if his bodily incarnation is laid bare by means of the thorn's aversive presence, making the flesh nailed to the self. There is on the one hand nothing that suggest that Paul thinks of the thorn as punishment for former sins, and thus the parallel to Job's torments is clear: both are stricken by Satan with physical pain and disease, yet no reference to violations of the moral order or the Law is in question. On the other hand, there is neither any suggestion of a mystic or romantic view of pain as a royal road to spiritual refinement—indeed, it meant the end of his mystical visions. There is no sign a stoic insensibility to its pain, nor promise of particular gains at the end of such suffering. Similar to Christ in Gethsemane, Paul asks three times to be spared of the torment (2. Cor. 12: 8). This conveys that the thorn in the flesh is not only unwanted; it is experienced as an evil invading the very bodily self that Paul cannot but want to get rid of. It hurts, makes him impotent and incapable; it overthrows the sway of the "I can," and puts Paul in the weak position of "I cannot."

Weakness and Passivity

In Paul, the thorn occasions an *epoché* by bracketing his normal body and opening up experiences of embodiment that were previously concealed. But note that this *epoché* does not lead back to the normal body of "I can." While uninvited, weakness and pain become for Paul ways to identify himself with the crucified. Paradoxically, the identification at the same time entails a sense of power from the risen one. However this power is manifest in the apostle's life, it emphatically does not mean lifting the suffering: "Three times I appealed to the Lord about this, that it would leave me," Paul reports, "but he said to me, 'My grace is sufficient for you, for power is made perfect in weakness'" (2. Cor. 12: 8–9). I have suggested earlier that Paul's rhetoric makes an intricate employment of boasting, in terms of weakness and power. Insisting on boasting—not of strength but of weakness—Paul in effect is undercutting any established sense of boasting. This time, weakness is reversed yet again, not by turning from strength to weakness, but from weakness to strength—but without restoring the body. Certainly, "my grace is sufficient for you" is a tough message in this context. But still, the final time Paul boasts, it is neither of human strength, nor of his own weakness, but of a strength at another level, one coming toward him, as a gift received from a source outside himself. Since decidedly not stemming from himself, the power can only be received in a state where Paul's own power and capabilities are inhibited. Hence, it is not as much a paradox, as a disclosure of a new asymmetry of weakness and power that enables Paul's experience to arrive: "So, I will boast all the more gladly of my weaknesses, so that the power of Christ may dwell in me" (2. Cor. 12: 9).

The source of the power (sometimes called grace), unsurprisingly, stems from the risen Christ. But apart from Paul's Christian conviction and conception—what, if anything, can this power phenomenologically denote? It cannot be investigated with the means of the usual understanding of the lived body (*Leib*), since that body is essentially body-intentionality that correlates with the world. But no worldly power is here at stake. Neither does it stem from the sources of the body itself, since it precisely arrives in the "I cannot," which the body undergoes deprived of powers. Nevertheless, the power is decisive to Paul's notion of the body, even if it does not heal the brokenness of his biological condition, as miracles might do. On the contrary, the sufficiency of grace only holds Paul firm in his bodily weakness and confirms it. And yet, the experience of the power in weakness might not be unreachable for phenomenology, if we refuse to take the lead back from brokenness to normal activity, but follow the lead further into the state of passivity of "I cannot," as the very thing itself.

Michel Henry proves relevant in this respect, arguing that there is a stratum—called life—that does spill over into perceptions and movements, activity and will, but which is only captured as such if one allows for a further, radical reduction. For life as it is incarnated in the human flesh is not, according to Henry, visible in the way that Husserl, Merleau-Ponty and Heidegger presuppose, say within the correlate of intentionality and world. Rather, life shows itself in what Henry denotes as our flesh, prior and independently of the lived body and visible world: in impressions pure and simple, in affects such as hunger and thirst, fatigue and vitality, pleasure and suffering. Unlike intentionality, which directs itself to a phenomenon at a distance, impressions and affects are always immediately given to themselves; there is no distance or distinction in being in pain or in pleasure—it is me undergoing my own affections, in one word: auto-affection. Impressions and affections are all auto-revealing without external reference or inner remainder.[34] The lived body must therefore be reduced to its immanent essence, to flesh, which is manifest in all impressions and affections. The constant presence and movements which unfold themselves in such flesh are the movements of life itself, according to Henry's conception. He writes: "Only absolute, transcendental, phenomenological Life—whose property is precisely revealing itself to itself in its pathos-filled self-affection, which owes nothing to anyone or anything else—can define human reality as phenomenological in its essence."[35] Life is indeed the ultra-phenomenon which reveals itself—which in the end is generated and identified with God.

I do not want to enter any discussion of Henry's notion of God here, but even if it might be one-sided, Henry points toward a dimension of the incarnated body which has been given little attention in the phenomenological literature, and which is crucial to my concerns here, namely its invisible and passive dimension.[36] Henry frequently refers to suffering and physical pain as cases in point, not in order to bring out their special phenomenological profile and quality, but to employ them as representative for impressions in general. When the self-revealing experience of pain is undergone by the immanent flesh, a deep powerlessness is manifest. As with Paul's thorn in the flesh, Henry refers to the fact that pain occurs beyond, even contrary to our will and power, and that we are incapable of ridding ourselves of it.[37] Interestingly, Henry does not confine himself to the analysis of bodily weakness and its affections, but proceeds to where the human power is obviously most patently at work: in our "I can."

Henry does not deny that human will and potency are at work in sensation, touch and movement, as pointed out by Husserl and Merleau-Ponty. Merleau-Ponty speaks of body image in which sensation and movement, indeed, all of my powers work together in the communication with the world. Henry, however, finds such analyses inadequate.[38] They are inadequate due to their constant reference to the correlation between perception and the world, and especially because they do not interrogate where the power of "I can" originally stems from. Henry contends that the mundane "I can" wells up from a more ancient power than the ego's, one that is given in and through life itself. Life does not well up from the center of the constituting ego, but it arrives in flesh, Henry will say, alluding to the incarnation of Christ. Hence, life is a power that enables every "I can"—seeing, hearing, touching, moving—a power over which we, in the final analysis, have no control: we cannot bring it about, nor rid ourselves of it, as it belongs to the gift of life itself. As Henry puts it: "*Every power collides in itself with that about which and against which it can do nothing, with an absolute non-power. Every power bears the stigmata of a radical powerlessness.*"[39]

The stigma of powerlessness inscribed in the "I can" returns us to Paul's stigma, the thorn in the flesh. I think Henry's location of an ultra-transcendental notion of life makes much sense as a phenomenological equivalent to Paul's speaking of grace and power: it arrives in me and becomes the essential part of me, remains invisible to mundane strength or weakness, and yet makes all of my abilities possible in the first place. For Henry, it is never a question of privileging the disclosing occasioned by the broken body, whereas Paul seems to insist on precisely this. Not only is Paul's thorn an intrusive *epoché* or reduction that turns him toward his bodily constitution and finitude, but, above that, it entails a kind of radical reduction, to the strata that prepares for the coming of power: weakness. There are both theological and phenomenological reasons for privileging the weak body because one of the permanent threats Pauline theology warns against is disowning God's gift. It has many expressions, but paradigmatically it is the works of the law: "I do not nullify the grace of God; for if justification comes through the law, then Christ died for nothing" (Gal. 2.21). Paul's point is that the cross has changed the view of the law in regard to salvation: the works of the law do not lead to salvation; they even work against it in so far as we take advantage of the law, drawing on our own powers to fulfill it and hence earn salvation by our own means. According to the latter self-image, there is no room for receiving; the self is already fulfilled by the means of the "I can." It is precisely this which is the problem with boasting: it invests value in powers and self-sufficiency to the degree that the broken body cannot but be regarded as lack or deviance. Its self-sufficiency closes off the passivity that makes the vital receptivity discernible. The insight made manifest in the reduction to weakness is precisely that we are constituted by receiving powers in a state of passivity.

There is a famous metaphor in which Paul brings home this point with clear relevance to the thorn in the flesh: "But we have this treasure in clay jars, so that it may be made clear that this extraordinary power belongs to God and does not come from us." (2. Cor. 4: 7). The metaphor of the clay jar, I take it, is meant to suggest a certain hollowness, which makes it possible to receive and contain the treasure in the first place. It robs the "I" of its "can" in the definition of the subject's ultimate status. The fragility of the clay jar parallels the way Paul in other places boasts of his weakness, since the value should not be tied to the vessel itself. The hollowness, and weakness, is a privileged human condition in so far as it makes the receptivity of Christ's power possible. Unlike strength, hollowness and weakness produce no illusion of where the treasure comes from: not from me but toward me. The jar is priceless, not in its own terms—say, my own resources and capacitates—but due to the treasure contained therein. This holds also for the view of the body: this view does not in any way exclude activity and achievement, but it entails that that any such activity is not drawn from the self, but enabled by powers received. Returning to my previous question: how does Paul's admitted weakness fit with the discourse of strength? As the image of the clay jar suggest, while the body is fragile and weak, the strength can come to it from outside its confines. The weaker the body, the more manifest is the strength: "power is made perfect in weakness" (2. Cor. 12: 9).

The Self as "I Cannot"

What kind of subjectivity is at stake here? What makes up our identity as incarnate human beings? It is true that the spontaneous ego of modern thought, right up to classical phenomenology, depends on an understanding of an ego shaping itself according to active achievements, notwithstanding the corresponding passivity within the facticity of our given surroundings. In taking the embodied self seriously, the decisive thing is how this phenomenology accounts for the body, or, more precisely, how its activity and passivity are conceived. If the embodied subjectivity realizes itself in activity, then "I can" makes up its essence, and "I cannot" turns out to be the self's limit—or even fall outside the limit of its conception. Breakdowns are important test cases for thinking of subjectivity, since they challenge phenomenology's conceptual ability to keep its integrity despite illness, weakness and dysfunctions. Subjectivity, according to classical phenomenology, fails this test, since it runs out of sound conceptual tools when exposed to "I cannot" due to the activist way constitution and identity are invested in its capabilities.

However, if phenomenology opts for another approach to selfhood and starts out, as Paul starts out, from the weakness of "I cannot," it can regressively dig into the power that nevertheless remains—life itself, call it grace, divine power. This does not abolish the relevance of the capable human being, and it is not a celebration of passivism in the sense of becoming passivized. Far from it. It digs into the roots of any activity of the capable subject. But the self's identity does not any longer depend on human strength and capability, which is why it is also manifest—indeed phenomenologically speaking, more pressingly manifest—in the weak, the "I cannot." Divested of all achievements that refer back to the self, in a state of passivity beyond the correlation between passive and active, the receiving of gifts becomes unavoidably clear. This state raises the question: "What do you have that you did not receive? And if you received it, why do you boast as if it were not a gift?" (1. Cor. 4: 8)

Ultimately, this entails not only an openness for receiving what is originally not one's own, but also a recasting of the understanding of subjectivity. Even as the received life and power do not originate in oneself, they make up the deepest layer of the self, thus subverting the priority of human activity and achievements. In Paul and Henry, the subject is always late, already conditioned by powers that precede it and keep it alive. This not only decenters the subject, but also entails another understanding of subjectivity, not only as the center from which intentional rays emanate, but as itself receiving itself. In one of Jean-Luc Marion's significant discussions of the self, he likewise suggests that the subjectivity is not originally the nominative sense of I, but in dative: the "onto whom," which means that it does not find itself at the point of origin, but as a receiver. "I receive *my self* from the call that gives me to myself before giving me anything whatsoever."[40] And precisely the weakness, brokenness, the thorn in the flesh, in short "I cannot," work as reductions opening us toward the receptivity of powers and life that stream to me and through me, as a gift: the incarnate self is the gifted.

Notes

1. Edmund Husserl, *Ideas Pertaining to a Pure Phenomenology and to a Phenomenological Philosophy. Vol I*, trans. F. Kersten (The Hague: Martinus Niejhoff, 1983), 44. For an elaboration of the carnal implication of such givenness in the flesh, see Didier Frank, *Flesh and Body: On the Phenomenology of Husserl*, trans. J. Rivera and S. Davidson (London: Bloomsbury Press, 2014), 18–23.
2. Maurice Merleau-Ponty, *Phenomenology of Perception*, trans. C. Smith (London: Routledge, 1962), 137; Edmund Husserl, *Ideas Pertaining to a Pure Phenomenology and to a Phenomenological Philosophy. Vol. II*, trans. R. Rojcewicz and A. Schuwer (Dordrecht: Kluwer Academic Publisher, 1989), 228.
3. Martin Heidegger, *Being and Time*, trans. D.J. Stambaugh (New York: SUNY Press, 2010), 309–15.
4. Merleau-Ponty, *Primacy of Perception and other Essays*, trans. and ed. J.M. Edie (Evanston, IL: Northwestern University Press, 1964), 14.
5. Merleau-Ponty, *Phenomenology*, 100.
6. Husserl, *Ideas II*, 252.
7. Ibid., 154; Merleau-Ponty, *Phenomenology*, 93. The ontological implications of the one hand touching the other is further developed in Merleau-Ponty, *Visible and Invisible*, trans. A. Lingis (Evanston, IL: Northewestern University Press, 1968), 139–41.
8. Husserl, *Ideas II*, 23–4.
9. Merleau-Ponty, *Phenomenology*, 152.
10. Husserl, *Ideas II*, 269.
11. Heidegger, *Being and Time*, 139.
12. Ibid., 252–3.
13. Ibid., 266.
14. Husserl, *Ideas II*, 346–7.
15. Maurice Merleau-Ponty, *Themes from the Lectures at Collége de France 1952–1960*, trans. J. O'Neill (Evanston, IL: Northwestern University Press, 1979), 27.
16. Heidegger, *Being and Time*, 135–7.
17. Havi Carel, *Illness: The Cry of the Flesh* (Durham: Acumen Publishing, 2013), 83.

18. Heidegger, *Being and Time*, 72–5.
19. Martin Heidegger, *Zollikoner Seminare*, ed. M. Boss (Frankfurt am Main: Vittorio Klostermann, 1987), 111.
20. Drew Leder, *The Absent Body* (Chicago: Chicago University Press, 1990), 71.
21. Merleau-Ponty, *Phenomenology*, 81–8.
22. Ibid., 136.
23. Ibid., 107–8.
24. Husserl, *Ideas* II, 270–1.
25. Ibid., 72.
26. Havi Carel, "The Philosophical Role of Illness," *Metaphilosophy* 45 (2014): 27.
27. Husserl, *Ideas II*, 266.
28. John A.T. Robinson, *The Body: A Study in Pauline Theology* (London: SCM Press, 1952), 9.
29. Hans Ruin argues that reversions and deconstructions of established hierarchies is central to his entire discourse. "Sacrificial Subjectivity: Faith and Interiorization of Cultic Practice in the Pauline Letters," in *Philosophy and the End of Sacrifice*, eds. P. Jackson and A.P. Sjödin (Sheffield: Equinox, 2016), 206. Georgio Agamben Speaks in Similar Vein About Paul's "Messianic Inversion." *The Time That Remains: A Commentary on the Letter to the Romans*, trans. P. Dailey (Stanford, CA: Stanford University Press, 2005), 97.
30. David Allan Black, *Paul, Apostle of Weakness: Astheneia and its Cognates in the Pauline Literature* (New York: Peter Lang, 1984), 139.
31. Cf. Candida R. Moss, "Christly Possession and Weakened Bodies: Reconsideration of the Function of Paul's Thorn in the Flesh," *Journal of Religion, Disability & Health* 16 (2012): 322–3.
32. Ronald Russell, "Redemptive Suffering and Paul's Thorn in the Flesh," *Journal of Evangelical Theological Society* 39 (1996): 565–6.
33. Cf. Gal 4: 13, where Paul refers to his disease in Gallia, possibly also in 1 Cor. 2: 3.
34. Michel Henry, *Incarnation. A Philosophy of the Flesh*, trans. K Hefty (Evanston, IL: Northwestern University Press, 2015), 59.
35. Ibid., 254.
36. The one-sidedness I have in mind is the excessive immanentism which makes it hard to see how this flesh can connect back to the world. For a criticism along these lines, see Renaud Barbaras, "The Essence of Life: Drive or Desire?" in *Michel Henry: The Affects of Thought*, eds. J. Hanson and M.R. Kelly (London: Bloomsbury Press, 2012), 52–3.
37. Michel Henry, *Incarnation*, 60.
38. Merleau-Ponty, *Phenomenology*, 98–101.
39. Michel Henry, *Incarnation*, 173 (Henry's italics).
40. Jean-Luc Marion, *Being Given. Toward a Phenomenology of Givenness*, trans. J.L. Kosky (Stanford, CA: Stanford University Press, 2002), 269.

2 Perceiving the Vulnerable Body
Merleau-Ponty's Contribution to Psychoanalyses

Ståle Finke

Introduction

In his *Phenomenology of Perception* from 1945,[1] Merleau-Ponty reports the case of a young woman who has lost her ability to sleep, her appetite and also her ability to speak as a result of her mother prohibiting her to see the man she loves. The case from 1929 is taken from Ludwig Binswanger's recording of his treatment of an Italian woman suffering from hysteria, a woman who was taken care of in the psychiatric clinic Bellevue in Switzerland where Binswanger was director. After two relatively short periods of psychoanalytic treatment in combination with more physical and also social interventions, the woman was released, apparently cured of her symptoms. According to Binswanger, the woman's self-relation was restored as the result of a changed *bodily* orientation, a different way of handling what he calls her "bodily self."[2]

For Merleau-Ponty, Binswanger's depiction supports his thesis that the clinical perspective of psychoanalysis and the phenomenology of the body are mutually supportive and enriching approaches to human experience: "It would be a mistake to believe that psychoanalysis . . . is opposed to the phenomenological method. Psychoanalysis has, on the contrary . . . contributed to developing the phenomenological method by claiming, as Freud puts it, that every human act 'has a sense.' "[3] In the light of this, several questions need to be addressed: how does the body make sense? And, how might the phenomenology of the body contribute to psychoanalysis? What is the role of the *unconscious*? And what might phenomenology learn from clinical experience?

Phenomenological thinking has proven particularly resistant toward the notion of the unconscious as this seems to introduce a depth to the mind hidden from the surface of intentional experience and conscious agency. For Merleau-Ponty, however, the unconscious belongs to our signifying bodies, to the ways in which bodily significations escape us, as we engage in an inter-corporeal and communicative field of sense-making.[4] The unconscious belongs to perceptual life, to our ways of perceiving and being perceived. In this manner, the unconscious is not something located in the inner realm of the psyche with causal relevance to current conduct and conscious expression; it is a *perceptual phenomenon* creating an "ambiguous presence . . . prior to every explicit recollection, like a field we open onto."[5]

In what follows, I hope to be able to explain what this claim amounts to. For Merleau-Ponty, clinical experience attains to a special idea of the body, the idea of the phenomenal body. In being exposed to one another, both the analyst and analysand participate in the mutual work of recovery and mourning, making the past and its dissociated significations into mutual responsibilities. On this basis, Merleau-Ponty is critical of the Freudian neutral interpreter, but does not deny the role of intervention and interpretation as in some recent psychoanalytic approaches.[6] In particular, in his later lectures on *Institution and Passivity* at the Collége de France from 1954 to 1955,[7] Merleau-Ponty addresses the problem of the relevance of interpretation in an interesting way. Clinical experience is an encounter of projection and misidentification, a process of mutual reversals, attunements, mis-recognitions and recognitions, where the distinction between participants is blurred and promiscuous. However, interpretation and articulation of past scenes of embodiment are continuous with the work of narrative recollection, where the analysand as well as the analyst come to discover the formative role of individual events, the very contingences of a unique biography.

Ultimately, this mutually enforcing idea of the clinical encounter and the phenomenal body points to a conception of self and subjectivity exposed to bodily vulnerabilities. The self is not the invariant subject of the philosopher, but a fragile sense of bodily selfhood exposed to failure, brokenness or pathology. According to Merleau-Ponty, the acknowledgment and articulation of bodily contingencies, of the bodily unconscious in speech and behavior, of our history of early and late dependencies, is not only intrinsic to a cure, but to our very conception of humanity. The phenomenal body as disclosed in clinical experience thus brings to expression our intrinsic sense of being bodily selves—having a history of past relatedness to be mourned and acknowledged. If the cure is this learning how to mourn, it is also enabling us to endure our sense of separateness.

The Body and the Structure of the Unconscious

Already in his early work, Merleau-Ponty advances a criticism of the theoretical constructions of psychoanalysis, confronting it with the idea of the body which might be articulated on the basis of clinical experience. This criticism, although somewhat modified in later writings, provides a particular instructive entry into Merleau-Ponty's thorough-going interest in the relation between phenomenology and psychoanalysis. Before we turn to the details of Merleau-Ponty's specific contributions, let us simply attempt to be clear about the context and aim of Merleau-Ponty's thought.[8]

In his *The Structure of Behaviour*[9] from 1942, Merleau-Ponty's remarks on psychoanalyses come as part of a general effort to integrate the contingencies of development into his conception of behavioral structure.[10] The psychoanalytic view of affective development is assigned a privileged status in this regard, due to its focus on the child's *affective* development.[11] However, in speaking of mental entities with causal significance, the Freudian meta-psychology is burdened with a classical mentalist or intellectualist picture of the mind. What we need to ask ourselves, Merleau-Ponty writes, is "whether . . . the psychological mechanisms which he [Freud] has described—the formations of complexes, repression, regression, resistance, transfer, compensation and sublimation—really require the system of causal notions by which he interprets them."[12]

Indeed, the Freudian topology stands in a long tradition of mentalist and Cartesian thought conceiving of the psyche as referring to private states discernible independently of the body and its engagements in the world.[13] Admittedly, in Freud's later writings, one finds an increasing skepticism toward the mentalist vocabulary; however, even in 1923, he still reveals himself incapable of thinking of affection and thought as originally *bodily significations*, stating that the psychic life of the ego somehow projects itself onto the surface of the body: "The ego is first and foremost a bodily ego . . . it . . . is itself the projection of a surface."[14] On this idea of the body, the ego is manifest on the surface of the body, still projected onto it; the unconscious, one would suspect, would be what retreats even from bodily visibility into a realm of inner psychic causation, irretrievable to personal reflection. What we ultimately are left with in Freud, then, is a picture of a doubled inner life of thought, affection and desire—mental significations or representations that are somehow conceived as self-sufficient entities, either as conscious states or as hidden significations, repressed thoughts and so on, haunting our emotional life. If we by contrast, as Merleau-Ponty suggests, think of our conscious life as embodied involvement with things and others, both consciousness and the unconscious will be aspects of bodily manifest conduct, features of lived presence.[15] On this idea of the body, of bodily presence in relation to others, the unconscious will, as Thomas Fuchs' recently has stated, represent a "horizontal dimension of lived space, most of all lodging in the inter-corporeality of dealings with others."[16]

The trouble in Freud's thought and research is exactly the confusion of two distinct aims, one of explaining conduct and expression in terms of hidden psychic entities and complexes, and one of clarification or elucidation of a certain *structure* or mode of conduct.[17] If one takes the clinical perspective into account, there seems to be a clear difference between saying that you are not aware of the true meaning of what you are saying and doing (as if the true sense was hidden from you due to some unconscious mental causes), and pointing out that there might be a discrepancy between what you say and think and your manifest bodily conduct. The body—posture, tone of voice etc.—betrays significations not thought of. The question is whether one needs to appeal, or to what extent (and in what context?) it makes sense to recur to intra-psychic entities or unconscious mental causes in order to *understand* behavior and the significations of what one says or feels. Indeed, the whole mentalistic appeal to intra-psychic entities seems to rely upon a false phenomenology of clinical experience, the idea of a neutral or disembodied interpreter inspecting the mind of another without him/herself being involved. As Freud recommended: "The doctor should be opaque to his patients, and, like a mirror, should show them nothing but what is shown to him."[18] However, as Binswanger made clear in a letter to the philosopher Erich Rotacher dated 14 June 1954, in clinical experience, bodily existence and mind are entirely enveloped, the self being undivided in its expression: "I am certainly a psychiatrist, and for that reason can still less than you, divide the 'soul' and 'body,' as I always have the undivided living human being before me."[19] It is in the mutual bodily presence to one another that signification is borne. As Merleau-Ponty writes in the *Phenomenology of Perception*: "[B]odily or carnal life and the psyche are in a reciprocal relation of *expression* . . . the bodily event always has a psychical *signification*."[20]

As soon as one discovers how clinical experience is founded upon inter-corporeal dealings of bodily significances, of relatedness and mutual implication, the very idea of a neutral observer or interpreter disclosing an intra-psychic realm of mental entities cannot even get started.[21] Clinical experience is a mutual work of affective sense-making, which implies an exposure of one's body to another. The unconscious is what comes to be expressed *within* this relation, and designates the background of former relations of embodiment. What is emerging from clinical experience is a certain phenomenological preponderance of an embodied encounter, where inter-corporeal communication is constitutive of the very appearing of affective sense, and where past relations of embodiment are expressed in current ones. I am present to the other only as a body, and the other body has significance in its presence to my own, being assumed, so to speak in my own posture, and in my relation to myself. Clinical experience, then, endorses an idea of the body different from Freud's: "[I]n contact with clinical experience, we see [in psychoanalysis] the emergence of a new idea of the body."[22]

This idea of the body is phenomenological and makes it impossible to distinguish psychic significations from bodily conduct. It means to take up the view of the phenomenologist for whom significations are expressed in conduct, as ways of being related, avoiding or defending oneself against such relatedness, longing to be contained by them or confirming them. When explanations are applied to the domain of sense, the result is often a reification of the patient's embodied sense of selfhood, underwriting and perhaps even fostering the patient's social isolation and withdrawal. Cure comes first when such resistance is overcome, when past relations of embodiment are exposed within a new setting, or, to use a term from recent developments in psychoanalytic thought, becomes *enacted* in the relation between the analyst and the analysand.[23]

As Merleau-Ponty writes:

> [T]he patient could not accept the sense of his disorder that has been revealed to him without the personal relationship that he has established with the doctor, without the confidence and friendship felt toward him, and without the change of existence that results from this friendship.[24]

The possibility of recovery, depends, in this way, upon a friendship capable of enduring and modulating a history of broken relatedness—of disappointments, neglects and withdrawals and traumas together with the resulting excesses of defensive and symptomatic strategies a self might have fallen prey to, and which are projected into the therapeutic situation. The notion of the unconscious that emerges from clinical experience is thus both bodily and relational—it reflects the duality of perceiving and being perceived. As such the unconscious is to be justified phenomenologically and not as an explanatory notion; it concerns the very bodily structure of *appearing sense*, a deep-structure of bodily intentionality and perception that reveals the presence of past relations, and the presence of the other, within current expression and conduct. The friendship established involves taking mutual responsibility for the past. The bodily field of unconscious significations is for the analyst and the analysand to share.

The Idea of the Perceiving Body

The idea of the body gained from a clinical perspective is the *lived* or phenomenal body, as opposed to the body as object in the world. It is the body of perceived significance, the body which takes a stance in the world, and toward others, in perceiving and being perceived. The body I am, and the body I perceive, the other, immediately convey signification—we are bodies that make sense, let things appear. By the same token, the perceiving and phenomenal body is intrinsically a vulnerable body exposed to the contingencies that emerge from being displayed in the world, opening up toward the environment, its own history of relatedness and to others. In order to recover this idea of the phenomenal body, however, we need to provide a somewhat more detailed presentation of Merleau-Ponty's idea of perception and of the perceiving-perceived duality.

The first dimension to the just-mentioned perceiving-perceived duality is the body being engaged in the sensible world—in seeing, hearing, touching, in exploring the surface of things. In *The Structure of Behaviour*, Merleau-Ponty makes his analysis clear by the example of the relation between the soccer player and her field. The engaged player on the field is not entertaining the field, nor the ball or other players as objects. Already a minimal depiction of engaged bodily conduct and perception serves to dismantle the mentalistic assumptions of classical philosophy: the player is not performing intellectual operations or conceptual inferences that are somehow applied to the field. Rather, the soccer field makes up an environment or *situation* for the body, which solicits certain preferred ways of conduct:

> For the player in action the football field is not an "object." . . . The field itself is not given to him, but present as the immanent term of his practical intentions; the player becomes one with it and feels the direction of the "goal," for example, just as immediately as the vertical and the horizontal planes of his own body.[25]

The player perceives the trajectory of the ball, where it will reach the turf, as well as where and how to hit it, not by some prior mental representations or calculation of the ball's whereabouts, but by being actively engaged in the ball's trajectory and behavior, adapting continuously to changing circumstances, continuously modulating eye-movement, posture and skill while keeping the ball within the range of a bodily grip. Certain movements are called for, so to speak, being pulled out of the player by solicitations from the environment—movements that hollow out significances to be acted out. Perceived significations are thus bodily contingent and extend into the environment. As Merleau-Ponty continues:

> [C]onsciousness is nothing other than the dialectic of milieu and action. Each manoeuvre undertaken by the player modifies the character of the field and establishes in it new lines of force in which the action in turn unfolds and is accomplished, again altering the phenomenal field.[26]

However, conduct is not restricted to a given situation; in enacting its surroundings, the body assumes *general* significations, it exhibits a *figurative* sense, invoking *possible* environments.[27] This means that a certain movement, a certain posture, a certain skill might be deployed again, attaining to a certain generality in its significance. Hence, the excellent player spontaneously deploys his/her general skills in novel situations, defining the situation, perhaps even the criteria of the game anew. Due to the complementarity of bodily skills and the perceived environment, the body is *displayed* in the world. Even the simplest perception thus points to its reversal: perceiving is not merely a touching of surfaces, but an ambiguous phenomenon of perceiving and being perceived.

The inherent duality of the perceiving-perceived is further subject to a detailed analysis in the *Phenomenology of Perception* which clearly reveals the intrinsic and latent complexity of the presumably simple beginning in bodily presence and perception. Perceiving a line corresponds to a certain posture or a way of being in the world—a line has existential significance, my body inhabits it, knows its attitude. The same holds for a color, the shape of a thing or its quality; the color blue, as Merleau-Ponty writes, is not some self-sufficient quality but a modality of the way the perceiving body explores a surface and retains a perceptual constancy throughout contingent circumstances of lighting:

> Blue is what solicits a certain way of looking from me, it is what allows itself to be palpated by a specific movement of my gaze. It is a certain field or atmosphere offered to the power of my eyes and of my entire body.[28]

The perceived thing is thus "shot through" with ambivalent significances, lived bodily significances and modalities of expression which recede into the depth of the thing: "The fully realized object . . . is shot through from all sides by an infinity of present gazes intersecting in depth."[29]

This already points forward to a fuller integration of the perceiving body in relations of mutual intercorporeality. Interestingly, the perceived phenomenon *as such*, the thing, implies the idea of an unconscious going to the depth of the perceived, spreading out into the environment to the extent that the thing has a history of significations, being something transitional in opening up toward symbolic articulation and re-articulation. This unconscious is bodily structured, being a field of intersection between my implicit bodily awareness and what I explicitly and intentionally attend to. In Merleau-Ponty's understanding, the bodily sense of the unconscious is thus a field of ambiguity or ambivalence of perceptual signification, of perceptual presence being undermined, being shot through with a history of significances, being hollowed out by gaps and fissures, fields of absence, even nonsense. The unconsciousness belongs to the very *Gestalt* of perceiving consciousness. "Freud," Merleau-Ponty writes "is on the point of discovering what other thinkers [i.e. the gestaltists] more appropriately named *ambiguous perception*."[30] The bodily unconscious belongs to any perception; it is the field of latent signification within the perceiving-perceived duality. What is more, it is this phenomenal fact which also requires that the very notion and scope of perception is extended beyond the mere perceiving of something present to include the body as being presently perceiving itself being perceived. In the final analysis, bodily significations are *intersubjective*, they signify relatedness, situations where perceiving entails being both displayed in an environment and being exposed to others, and to a history of relatedness.

This should also serve to dismantle the often raised objection to phenomenological reconstructions of the unconscious that they merely align
it to the notion of implicit consciousness, that is, the pre-conscious in the
Freudian sense.[31] However, as has just been suggested, and in contrast to
some phenomenologists, Merleau-Ponty does not terminate his analysis with the simple reference to perceiving consciousness and its implicit
horizon or background. Being perceived, the body is exposed to an other.
This perceived relatedness is sedimented in the bodily posture (the postural schema), and in the ways the body attends to, and invokes, the
presence of others. As there is a depth to things, there is a depth to our
embodied past with others. Thus I might meaningfully ask the question,
as does psychoanalysis, whose voice, attitude or affective complex I am
in fact voicing – speaking, gesturing or signifying, so to speak, beyond
myself. The work of clinical experience is to unfold and work through
the latencies and past significations in present talk, disclosing how affective complexes are voiced, and how the voicing of emotions reveals attitudes that bear upon unacknowledged meanings. The idea of the body
that emerges from clinical experience is thus the phenomenal body as it
is enmeshed in a history of inter-corporeal relations and significations.
One of the most important contributions of the analytic process is to
discover and sort out whose voice is currently voiced or projected in
order to recover a more adequate sense of self-reliance. And perhaps the
most important contribution of psychoanalysis as such is its discovery of
ambivalent significations – how my own voice is reversed by the language
of my body, being the matrice of former relatedness.

Perceiving the Body of the Other

In particular, what has, after Donald Winnicot, been called *object-
relation theory* implies a transformation of Freud's thinking that conceives the significance of the other to be intrinsic to the development
of the self.[32] Merleau-Ponty's contribution to this developmental story
concerns the way he founds identifications and early bindings, in love,
mutuality, attunement and recognition, constituting the way in which
one's embodied self is formed through being attuned and mirrored, that
is, perceived, in others: "It is not because the child has the same blood
as his parents that he loves them; it is because he . . . sees them turned
toward him, and thus identifies himself with them, conceives of himself
in their image, and conceives of them in his image."[33]

Already in early phases of development, the human baby typically
responds imitatively to the gestures and facial expression of primary caretakers, invoking the presence of others with the expectation of confirmation independently of their actual presence.[34] As Merleau-Ponty states in
his lectures at Collége de France from 1953: "[T]he child immediately
grasps the other's body as a carrier of structured behaviours and [. . .]

he experiences his own body as a permanent and global power of realizing gestures endowed with a certain meaning."[35] What Merleau-Ponty
was referring to early on is the phenomenon which later developmental
psychologist have called *primary imitation*—the infant's ability to imitate
the perceived bodily gestures of primary caretakers—suggesting a deep
affective correspondence between the infants' bodily awareness, the very
formation of their body schema, and the perceived conduct and bodily
presence of the other.[36] The affective presence of the other is perceived on
the surface of the body, in the form of a gesture, a smile, fatigue, anger,
frustration and so on—features that are simultaneously being fleshed out
in a perceiving relation, echoed in the infant's own sense of what Winnicott called their "psycho-somatic existence."[37] As Merleau-Ponty writes:
"The constant regulation of bodily equilibrium, without which no function would be possible in the child . . ., is not merely the capacity to reu-
nite the minimal conditions for balancing the body, but is more generally
the power I have to realize with my body the gestures that are analogous
to those I see."[38] Thus, the perceiving baby slips into the very shape or
body she perceives. At this stage, there is a promiscuous relation between
self and other, a form of affective symbiotic co-presence. From this stage
of intrinsic dependency and *holding*, in Winnicott's terms,[39] a self gradually emerges. This self is an embodied self, regulated and modulated in
virtue of being perceived.

The developmental significance of primary imitation has been richly
studied and confirmed in recent years, also affecting the psychoanalytic
understanding of clinical experience.[40] The means of conceiving, articulating and making clinically useful such things as transference, projection, introjection, counter-transference and so on, have undergone major
revisions in accordance with object-relation theory and the so-called
relational view. Melanie Klein's initial understanding of the notion of
projective identification, which has been of immense influence for this
development, is exactly aimed at grasping the promiscuous ways in which
a subject dispatches his/her conflicts, anxieties or affective tensions in an
other, and, thus, points to, in Klein's formulation, a situation where "[t]he
object to a certain extent becomes representative for the ego."[41] What is
radical in Merleau-Ponty's embodied phenomenology of primary relatedness and imitation is that there is no longer any need to entertain the idea
that such projection takes place as a transference from one inner psychic
self to another, but rather that projection and introjection are themselves
primary forms of bodily relatedness. As Merleau-Ponty writes: "The psychological mechanisms of introjection and projection, instead of appearing as spiritual operations, must be comprised as modalities of the body's
activity."[42] Projection and introjection are derived forms of the bodily
capacity for imitative behavior, where my own gestures are *transitive* in
finding their perceptual completion in that of others, and, conversely,
where I perceive the significance of those of others by being attuned to

them, re-enacting them: "Transitivism is . . . the same notion that psychoanalyst are using when they speak of projection, just as mimesis is the equivalent of introjection."[43]

Transitivism characterizes all stages of development—from the infant's first attunement with others and its surroundings, and the child's attempt to find his/her own pleasures, joys or griefs elaborated in the faces of primary caretakers, to the adult's finding of signs of his/her own depression or disease in the behavior of others. The eventual mature ability to perceive the confirmations of the other as recognitions of one's individual separateness also depends upon perceiving oneself being perceived in an other, and to acknowledge this as such. The different forms of transitivism are being played out in the therapeutic setting—so that, when being attended to, the very process of affective bodily communication, as well as speech, submits both analysand and analyst to the turmoil of mutual reversals and displacements. An early experience of shame might lead to a withdrawal from sociality and a fundamental inhibition that, in relation to an other, might be expressed as rejection and sense of sovereignty, displacing the shame in the other. Yet, it is one's self in the other that is attacked, in the form of a reversed self-contempt.[44] The unconscious is this field of ambivalent bodily and affective signification that emerges within relatedness. It is the mutual enactment of self-other displacements, something which implies that everything we say and do is of intrinsic ambivalence, on the edge of being reversed: "We should prefer this notion of ambivalence, which paints perfectly all that is equivocal in certain behaviours . . . attitudes of hate that are at the same time love, desires that express themselves as agony, and so forth."[45]

To the love and confirmation of the other there will always also be a corresponding anxiety or threat, which one will displace by a certain disposition of and readiness for defensive conduct and thought, aiming at splitting, neglecting or undoing an intolerable ambivalence. Melanie Klein speaks in the context of the infant's world metaphorically of the "good breast" and the "bad breast" and an initial "fear of persecution."[46] The truth to these metaphors, apart from their bodily air, is the duality they designate, the intrinsic ambivalence, which belongs to being exposed in the relation of perceiving and being perceived, being bodily modalities of generosity and openness or withdrawal and rejection. The other might mean perceptual completion and love, and thus provide the answer to our fundamental desire or longing and prompt generosity. But the other might also mean a displacement of self; the other's gaze might turn into an aggressive intruder under certain circumstances, which results in introjected fear, persecution and anxiety at being exposed. The splitting off of the two, of goodness and fearfulness, might indeed be regarded as an active and primary defensive attitude toward this intrinsic ambivalence. Inhibition or self-enclosure might be seen as a defensive way to avoid the threat of invasion, and thus the fear of complete exposedness.

Of course, such concepts are themselves ambivalent and fragile, but need to be conceived as modalities of the idea of the body gained from clinical experience, where the reversals of affections and expression, in transference and counter-transference are the very phenomena attended to.

As long as early relatedness is secured within the constancy and durability of primary bonds of imitative sense-making, the child will embark on development with a fundamental sense of being contained in the other (or being *held*, as Winnicott would say), the other being the extension of oneself, one's bodily presence in the world. Self-reliance is thus a result of a "good-enough" holding, a caring and acknowledging which survives the exposures to destructiveness and aggressive forms of both manic defensiveness and excessive projection, and thus makes the self capable of enduring ambivalence without withdrawing. It remains to be seen, however, how this is played out in the therapeutic setting, and also how Merleau-Ponty's contribution thus far can be made fruitful for the understanding and conceptualization of clinical experience.

The Role of Interpretation

The idea of the body that we have gained from the mutually enlightening analysis of clinical experience and Merleau-Ponty's notion of the phenomenal body is opposed to the idea of the repressed—and of the unconscious—as somehow inner causes or givens. From a clinical perspective, it is thus also suggested that we drop the notion of the analyst as a neutral interpreter, and instead see her role in affective co-enactment with the analysand. Indeed, clinical experience and its communicative situation must be conceived in view of the general idea of the phenomenal body as intrinsically enacting forms of (past) relatedness. But what does this leave for interpretation? In his *Structure of Behaviour*, Merleau-Ponty seems to reserve only a minimal and heuristic role to narrative recollection: "[T]he childhood memory and . . . the traumatic event which provides the key to an attitude . . . are not . . . the causes of the behaviour. They are the means of the analyst of understanding a present structure or attitude."[47]

Of course, saying that a traumatic event is not the cause of the symptom is not to say that, e.g. defensive neurotic behavior, is somehow a deliberate matter. In the light of what has been said thus far, it is only to say that the trauma or unbearable situation in question is only *available* to us in terms of the lived significations that are expressed and structured by conduct and inter-corporeal communication with others. Merleau-Ponty thus seems to support certain current clinical developments where the role of the interpreting analyst, as Jessica Benjamin makes clear, is replaced by "a relational view that sees the importance of mutuality within the relationship—such that the mutually created dynamic is seen not only as an object of knowledge, but also as a source of transformation through regulation and recognition."[48]

There might, however, be a clinical sense of interpretation which is not assuming a distant observer's somehow cognitive access to causes, but rather conceives interventions to open up and articulate, through narrative recounting, the contingent and unique circumstances of events signified in behavior and in defensive conduct, thus opening up for alternative forms of mourning, modulation and re-working *within* forms of mutual recognition. Let us briefly return to Merleau-Ponty's discussion of Binswanger's case. The Italian woman in Binswanger's treatment suffered from aphonia which was, presumably, acquired in early childhood facing the trauma of an earthquake. At the age of 18 the experience of the earthquake is re-invoked, however falsely, in the wake of which she loses her voice. The mother's later prohibition to let her see her beloved comes at the age of 24, where loss of speech is accompanied by loss of appetite among other symptoms.[49] The mother's prohibition thus re-invokes the physiognomy of the childhood experience, and prompts her falling back upon habituated patterns of defensive conduct. As Merleau-Ponty writes:

> In the subject's childhood, fear was expressed by aphonia because the imminence of death violently interrupted coexistence and reduced the subject to her own personal lot. The symptom of aphonia reappears because the maternal prohibition brings back, *figuratively* [my emphasis], the same situation.[50]

In other words, the woman's problem, as Merleau-Ponty sees it, is how the loss is *figured* within a history of relatedness, and how her defensive conduct constitutes a general ability or disability solicited in seemingly diverse situations. Certainly, simply bringing the woman to acknowledge the role of the event of the earthquake will achieve very little since her problem concerns a general defensive form of corporeal relatedness, which has become pathological or excessive as it undermines communicative sense and narrows down her field of inter-corporeal communication. The cure does not consist in working things out at a cognitive level, but is to be found in novel ways of re-enacting the affective scenario in the co-presence of an other (the analyst):[51] "We need an other who knows without saying, or at least is ready to know, taking into account countertransference, who marks the place of truth which is not in the *struggle*, but in the co-presence."[52] In other words, the therapeutic process is the playing out of mutual enactments and reversals in a secure and holding environment, a transitional or *potential* space in Winnicott's terms, leading toward a display and modulation of one's own defensive attitudes in the presence of the body of an other.[53]

For at least two reasons, the clinical situation as just depicted encompasses interpretation, if understood adequately, rather than expels it. To begin with, psychoanalysis is not merely content with the building of a novel friendship, but is also dealing with symptoms and their *genesis* in order to improve the patient's self-understanding. Thus, As Daniel Stern makes clear: "One of the major tasks of the therapist is to help the patient to find a narrative point of origin."[54] It is for instance crucial to any adequate understanding of particular symptoms to be able to determine in what developmental phase a trauma or event has taken place, in order to understand the nature of the symptom.[55] Accordingly, even if one discards the Freudian schema of levels of sexual development and replaces it with the relational stages of self-other perception as in object-relation theory and beyond, it will still make a difference whether or not the trauma concerns the body directly (such as the Italian woman's trauma of an earthquake), and when a trauma occurred (say in a pre-linguistic phase prior to the relative independence from caretakers or after the entry into language, as suggested by the woman's aphonia). Second, and of equal importance, the very process of mourning may depend crucially upon an improved self-understanding and the ability to articulate this in interpretation. An interpretation in this light is the explicit taking over of responsibility in accepting loss, and, thus, opening oneself up for a mutual *understanding* and, thus, recognition in Jessica Benjamin's sense, of an individual and contingent biography. Interpretation, rather than being the stance of a disembodied analyst, opens up the potential space of mourning, leading finally or hopefully to the possibility of mutual separation and a sense of recovery through the resulting self-reliance. Acknowledging separateness is a mutual achievement where the other is no longer merely used or projected, but generously accepted in her difference.

Trauma and the Recovery of Voice

In his lecture-course *Institution and Passivity* at the Collège de France from 1954 to 1955, Merleau-Ponty seems to suggest a more positive role for interpretation: "If the unconscious is not a thinker in us who knows everything, and not the event-based genesis of our 'symbolic matrices'— nevertheless we cannot reduce it to these symbolic matrices . . ., since analyses liberate concrete recollections which appear to be absolutely forgotten."[56] This points forward and attempts at a reintegration of interpretation within the clinical context: an adequate phenomenological understanding of the clinical perspective needs not only to account for the generality of dissociated conduct, but also for the latent significations of this conduct, that is, the genesis of symptomatic behavior in the contingent history of the patient. This can only be done in recollection, in finding a narrative "point of origin", to use Stern's phrase.

However, does not this return to interpretation yield a truthfulness to past events that conflicts with our analysis so far? On Merleau-Ponty's account in his lectures, though, an event no longer has causal, but only *institutional* significance. An event founds a norm or a style of conduct.

In this way, a trauma presents us with significances that are yet to be lived, modulated within relations of inter-corporeal exchange, of recognition and misrecognition. The event premonitions a certain *intersubjective* meaning, as it has taken hold of a bodily symbolic matrice in Merleau-Ponty's sense. The body attains to a language. The repressed is thus to be considered as a *dislocated* signification, a signification for which there might be no ordinary (or at least explicit) memory; it presents us merely with gaps and fissures, a sense of discontinuity between our self-understanding and the language of our bodies. The event or original scene is thus never at any stage truly original; it is exactly a phenomenon of great ambivalence, presenting us with a lack, lacunas of sense or "regions of non-sense,"[57] which are working upon us only by being severed from conscious acknowledgment in our efforts at making ourselves understood or contained by others. As Merleau-Ponty writes: "[T]he inaccessible installs itself as a norm and takes possession of our body, it dictates our body's movements precisely because it has been . . . repressed."[58]

The aphonia of Binswanger's woman is her way of refusing inter-corporal transactions, which, presumably, are experienced as threatening, invoking the physiognomy of an early situation. The repressed content, then, has no determinate signification; it presents only the way in which her present perceptual and bodily awareness is shot through by traces of a past, and with significations of past relatedness yet to be explored. And it is this complex of relatedness, and the defensive conduct that has colonized it that bears the significance of the event. The earthquake is the narrative point of origin that allows us, and her, to see a general pattern. It bears on no truth apart from this. Moreover, it is clear that her recovery is not the result of a cognitive effort, but of a living-through of an embodied past in relation to an other. As Merleau-Ponty remarkably puts it: "The patient will discover her voice, not through an intellectual effort . . . but . . . when the body again opens to others or to the past, when it allows itself to be shot through by coexistence and when it again signifies (in the active sense) beyond itself."[59]

Interpretation and recollection, then, are not somehow concerned with the truth or truthfulness or a certain memory-report, but reflects the *disclosure* of an overall physiognomy, pointing to past scenes of embodiment. Interpretation is an articulation of a past, which has become a present, which is currently co-enacted by analysand and analyst. It is an *indexical* invocation of a former embodied situation that makes its significance intersubjective, and thus something to be modulated and regulated within the new relation. As Merleau-Ponty indicates: "The unconscious is the symbolic matrix left behind by the event. The return to the event, analysis, *Deutung*, unravels this fabric, but these are only effective if that which generates the event is truly recovered as it was lived."[60]

In displaying the institutional significance of a trauma, both analyst and analysand are *in truth*: "The analyst and the analysand [are] both in the truth . . . as *aletheia*, unconcealment."[61] The Greek connotation invoked by Merleau-Ponty is that of bringing something forth from concealment. The concealed significance is displaced by being *figured* in conduct, and in defensive forms of relatedness. The new mode of relatedness is un-concealing relatedness, exposing the body to others, to its past, being shot through with significances to be acted out. This requires a new sense of separateness and a transitional and symbolic space in which to work this out in mutual recognition. In interpretation this space is articulated, disclosed as a space of mourning and recovery as well as a space of mutual understanding. Retrieving one's voice is thus something granted by the other, yet only as a form of accepting loss, of enduring ambivalence and the conflicting and turbulent contingencies of being dependent and thus vulnerable.

Conclusion: The Ethics of the Vulnerable Body

On Merleau-Ponty's account of clinical experience, the analytic setting is the staging or playing out of situations of primary embodiment—the intricate ways in which we are related through imitative behavior and the way in which this sediments bodily and affective significances in early as well as later phases of development. The idea of the body that emerges from psychoanalysis is that of the vulnerable body exposed to others in relations of perceiving and being perceived. Rather than conceiving of the body as an expression of the psyche, clinical experience has it the other way around; the psyche is an expression or manifestation of the body in inter-corporeal relations to others and is the result of the affective development of intrinsic forms of relatedness, the perceiving of the self in the other. In the clinical setting as in ordinary life, one must risk mutual reversals and the event of misrecognition, which in the end, if the analysand and analyst are both preserved throughout the sessions, serve to disclose a common understanding of symptomatic behavior, and a mutual openness for, and responsibility toward, the past. Such openness, being the result of friendship and acceptance, but also of successful interventions and interpretations, might prompt the analysand's conversion to other less defensive ways of modulating her affective dependencies in awareness of the other's independence, and also, finally of her own.

Depicted this way, psychoanalysis is already a form of embodied existential phenomenology—the *playing* out of the inter-corporeal forms of projective identification and introjection, of negative and positive reversals, of transference and counter-transference, of mis-recognitions and the return to, and acknowledgment of, separateness.[62] Achieving separateness and one's own voice is thus a major aim of the psychoanalytic cure—becoming less susceptible to the fluctuations of the early stages of imitative conduct, of being invaded by—and invading—others, and setting up defenses toward this threat of alienation. Interpretation opens up the symbolic space of mutual recognition where lived significances are expressed and articulated in language, hence opening up a space of mutuality where the other is not used projectively, but, say, projected more usefully as her own separate being, to be acknowledged, not destroyed or devalued.[63]

In his essay "Man and Adversity" from 1951, Merleau-Ponty conceives the contribution of psychoanalysis and its idea of the body to consist in its way of making us aware of our contingency and bodily vulnerability. A genuine humanism, Merleau-Ponty writes, "begins by becoming aware of contingency. It is the continued confirmation of an astonishing junction between . . . my body and myself, my self and others, . . . violence and truth."[64] The clinical situation embraces both—violence or loss as well as truth. This seems to forward an existential claim concerning our human condition. As Judith Butler has put it: "Loss and vulnerability seem to follow from our being socially constituted bodies, attached to others, at risk of losing attachments, exposed to others, at risk of violence by virtue of that exposure."[65] Psychoanalysis for Merleau-Ponty is a way to learn to accept being bodies, and, thus, to acknowledge vulnerability as intrinsic to our sense of self. The idea of the embodied self emerging from clinical experience is thus ultimately ethical. I am vulnerable only in perceiving the vulnerable bodies of others and in learning how to endure our separateness, coming, hopefully, to renounce upon the transgression or violence that results from our desire to transcend ourselves. If this promises an ethics, this would be an ethics of intrinsic ambiguity.

Notes

1. Maurice Merleau-Ponty, *Phenomenology of Perception*, trans. Donald A. Landes (London, Routledge, 2012).
2. Ludwig Binswanger, "Über Psychotherapie," in *Ausgewählte Werke Band 3, Vorträge und Aufsätze*, eds. Max Herzog (Heidelberg: Roland Asanger Verlag, 1994), 230. For a critical review of Binswanger's case, see Susan Lanzoni, "Existential Encounter in the Asylum: Ludwig Binswanger's 1935 case of Hysteria," in *History of Psychiatry*, 15, no. 3, 285–304. Binswanger's case must have appeared instructive to Merleau-Ponty for personal as well as for scholarly reasons. In his young days, Merleau-Ponty was himself suffering from having to comply with the wish of the parents of his beloved one, Elisabeth Lacoin, to withdraw from her and to pretend he was no longer in love. Consequently, he refused to see her and responded only ambivalently to her letters, something which led to her sudden breakdown, and, finally, also to her death. See Claude Francis and Ferdinande Gontier, *Simone de Beauvoir: A Life, A Love Story* (London: Vermilion Books, 1988), chapter 3.
3. Merleau-Ponty, *Phenomenology of Perception*, 160–1.
4. Ibid., 148.
5. Ibid., 381.
6. For an interesting discussion of these developments, see Jessica Benjamin, "Where's the Gap and What's the Difference? The Relational view of Intersubjectivity, Multiple Selves, and Enactments," *Contemporary Psychoanalysis*

46, no. 1 (2010): 112–19. Apart from a few exceptions, Merleau-Ponty's phenomenology of the body has not been given much attention in the psychoanalytic literature. See e.g. Roger Brooke, "Merleau-Ponty's Conception of the Unconscious," *African Journal of Psychology* 16 (1986): 126–30. More recently, however, there is an increasing literature and research, also empirically, where Merleu-Ponty's thought is applied to clinical issues and to conceptualizations of pathological conditions. See e.g. Thomas Fuchs, "Corporealized and Disembodied Minds—A Phenomenological View of the Body in Melancholia and Schizophrenia," *Philosophy, Psychiatry and Psychology* 12 (2005): 95–107.

7. Maurice Merleau-Ponty, *Institution and Passivity—Course Notes from the Collège de France (1954–1955)* (Evanston, IL: Northwestern University Press, 2016).
8. I will not in this article take the whole scope of Merleau-Ponty's texts and developments into account, and, in particular, I will not address the place of psychoanalysis in his last, but unfinished *The Visible and the Invisible* (Evanston, IL: Northwestern University Press, 1968). For a survey of these developments see e.g. James Phillips, "Merleau-Ponty's Non-Verbal Unconscious," in *Unconsciousness Between Phenomenology and Psychoanalysis (Contributions to Phenomenology 88)*, ed. Dorothée Legrand and Dylan Trigg (Cham: Springer Verlag, 2017), 75–95. My focus here is Merleau-Ponty's phenomenology of the body as perceived from the clinical perspective in psychoanalyses, and how this might contribute to recent conceptualizations within this field. In this regard, Merleau-Ponty's early work, as well as in some of the lecture-courses, both at Sorbonne as well at Collége the France, are clearly more pertinent as they explicitly address the questions taken up in clinical discourse.
9. Maurice Merleau-Ponty, *The Structure of Behaviour*, trans. Alden L. Fisher (Pittsburgh, PA: Duquesne University Press, 2008).
10. In *The Phenomenology of Perception* the discussion of psychoanalyses concerns the relevance of desire for a phenomenology of the body, and how to avoid thinking of this as a primary form of sexual being, but rather as a bodily ambivalent structure penetrating perceptual life as a whole. See *Phenomenology of Perception*, 156 ff.
11. Merleau-Ponty, *The Structure of Behavior*, 176–7.
12. Ibid., 177.
13. In this way, one might say Freud still subscribes to an introspectivist picture of the mind, even if doubled in conscious and unconscious psychic significations, and despite his effort to undertake what Paul Ricoeur called an "archaeology of the subject." Paul Ricoeur, *The Conflict of Interpretations*, ed. John Ihde (New York and London: Continuum, 1989), 158.
14. Sigmund Freud, "The Ego and the Id," (1923) in *The Essentials of Psycho-Analysis*, ed. Anna Freud (London: Vintage Books, 2005), 451.
15. See also Robert D. Romanyshyn, "Phenomenology and Psychoanalysis," *Psychoanalytic Review* 64 (1977): 211–23.
16. Thomas Fuchs, "Body Memory and the Unconscious," in *Founding Psychoanalyses Phenomenologically, Phenomenologica 199*, ed. Dieter Lohmar and Jagna Brudzínska (Dordrecht: Springer Verlag, 2012), 82.
17. Similar criticisms had already been developed not only in within phenomenology, but also by certain representatives of philosophy of language, such as Ludwig Wittgenstein. Wittgenstein typically addressed the discrepancy between "manifest" and "latent content," so as to object to any latent content being merely unavailable and of causal significance, and thus, conceived independently of the expressive and behavioral circumstances of an

expression, see Frank Cioffi, *Wittgenstein on Freud and Frazier* (Cambrdige: Cambridge University Press, 1998), 218.

18. Sigmund Freud, "Recommendations to Physicians Practicing Psycho-Analysis," (1912) in *The Standard Edition of the Complete Psychological Works of Sigmund Freud XII*, ed. James Strachey (London: Vintage Books, 1999), 114.
19. Quoted in Lanzoni, "Existential," 292.
20. Merleau-Ponty, *Phenomenology of Perception*, 162–3.
21. As Thomas Fuchs makes clear: "A therapist . . . who regards himself only as a projection screen would be in danger of missing the dimension of genuine encounter where he is met as a real, embodied person." "Psychotherapy of the Lived Space: A Phenomenological and Ecological Concept," *American Journal of Psychotherapy* 61, no. 4 (2007): 434.
22. Maurice Merleau-Ponty, *Signes* (Evanston, IL: Northwestern University Press, 1964), 228.
23. See also Ibid., 229. In the psychoanalytic literature this notion of "enactment" refers to the mutual and relational exchange of affective dispositions, and in particular in relations of transference, counter-transference, projection and introjection as played out within the therapeutic setting. As such it is thus intended to capture the relatedness and co-implication of self and other, and concerns how an active relation to such enactments can be made productive for clinical development and cure. See in particular Lewis Aron, "The Paradoxical Place of Enactment in Psychoanalysis: Introduction," *Psychoanalytic Dialogues* 13, no. 5 (2003): 623–31. See also Maria Ponsi, "The Evolution of Psychoanalytic Thought: Acting out and Enactment," *The Italian Psychoanalytic Annual* (2013): 16. The clinical notion approaches the use of the term "enactment" as it has been recently employed in philosophy and cognitive science, referring to how perceptual sense, and meaning more generally, depends upon the resources of bodily movement, that is, a skill-based mastery of sensorimotor contingencies. See Alva Noë, *Action in Perception* (London: The MIT Press, 2004). This perspective is very much in line with Merleau-Ponty thinking. However, so-called enactivism within cognitive science and philosophy has thus far entertained a rather limited idea of the body, and neglected the affective scope of enactments as thematized in the psychoanalytic literature. But, within what Shaun Gallagher has coined as a "broad enactivism", the psychoanalytic perspectives should be more than welcome. See Shaun Gallagher, *Enactivist Interventions* (Oxford: Oxford University Press, 2017).
24. Merleau-Ponty, *Phenomenology of Perception*, 166.
25. Merleau-Ponty, *The Structure of Behaviour*, 168.
26. Ibid., 169.
27. Interestingly, in order to grasp the behavioral dimension of meaning as outlined in the mutual interaction of body and environment, Merleau-Ponty adopts a term from Kurt Goldstein, who treats meaning as the result of a certain generality of bodily conduct and expression, what he calls *categorical behaviour*, that is, "the ability to deal with that which is not real—with the possible . . . the capacity to experience and to handle 'meaning.' " Kurt Goldstein, *The Organism* (New York: Zone Books, 2000), 44–5.
28. Merleau-Ponty, *Phenomenology of Perception*, 218.
29. Ibid., 71.
30. Merleau-Ponty, *Signes*, 229.
31. Struggling with conceptualizing the unconscious phenomenologically Rudolph Bernet has suggested that it might simply be regarded as the self-forgetting of the implicit and horizontal features of intentional consciousness: "the unconscious is the self-forgetting abandonment to perceived reality",

"Unconscious Consciousness in Husserl and Freud," in *The New Husserl: A Critical Reader*, ed. Donn Welton (Bloomington and Indianapolis: Indiana University Press, 2003), 217. A similar approach is taken in Dermot Moran, "Husserl's Layered Concept of the Human Person: Conscious and Unconscious," in *Unconsciousness Between Phenomenology and Psychoanalysis (Contributions to Phenomenology 88)*, ed. Dorothée Legrand and Dylan Trigg (Cham: Springer Verlag, 2017), 3–25. These approaches forget that the unconscious is *discontinuous* with present experience, and, thus, cannot be conceived as the mere implicit background neglected or blended out in intentionally directed experience.

32. See e.g. Donald W. Winnicott, "The Theory of the Parent-Infant Relationship," *The International Journal of Psychoanalysis* 41 (1960): 585–95.
33. Merleau-Ponty, *Signes*, 227.
34. Merleau-Ponty, *The Structure of Behaviour*, 166.
35. Maurice Merleau-Ponty, *Child Psychology and Pedagogy—The Sorbonne Lectures 1949–1952* (Evanston, IL: Northwestern University Press, 2010), 23–4.
36. For a discussion of the recent psychological research and literature on this, see Shaun Gallagher, *How the Body Shapes the Mind* (Oxford: Oxford University Press, 2005), 65 ff.
37. Winnicott, "The Theory of the Parent-Infant Relationship," 590.
38. Merleau-Ponty, *Primacy of Perception*, 145.
39. Winnicott, "The Theory of the Parent-Infant Relationship," 590.
40. See Daniel N. Stern, *The Interpersonal World of the Infant* (London: Karnac Books, 1985). For a synopsis, see also Nancy McWilliams, *Psychoanalytic Diagnosis* (New York: The Guilford Press, 2011), 31 ff.
41. Melanie Klein, *Envy and Gratitude and Other Works 1946–1963* (London: Vintage Books, 1997), 68–9.
42. Merleau-Ponty, *Child Psychology and Pedagogy*, 286.
43. Merleau-Ponty, *The Primacy of Perception* (Evanston, IL: Northwestern University Press, 1964), 148.
44. See the very interesting study by Lauren Levine, "Into Thin Air: The Co-Construction of Shame, Recognition and Creativity in an Analytic Process," *Psychoanalytic Dialogues* 22 (2012): 456–71.
45. Merleau-Ponty, *Child Psychology and Pedagogy*, 73.
46. Melanie Klein, *Envy and Gratitude*, 176 ff.
47. Merleau-Ponty, *The Structure of Behavior*, 178.
48. Benjamin, "Where's the Gap and what's the Difference?" 113.
49. Binswanger, "Über Psychotherapie," 213.
50. Merleau-Ponty, *Phenomenology of Perception*, 163.
51. As John Russon has put it, one would need to guide the woman "not to act in specific ways simply because they are familiar, but to become familiar with the specific ways she wants to act." *Human Experience* (Albany: SUNY Press, 2003), 88.
52. Merleau-Ponty, *Institution and Passivity*, 119.
53. Cf. Donald W. Winnicott, *Playing and Reality* (London: Routledge, 1971), 144 ff.
54. Stern, *The Interpersonal World of the Infant*, 262.
55. Ibid., 258 ff.
56. Merleau-Ponty, *Institution and Passivity*, 204.
57. Merleau-Ponty, *Phenomenology of Perception*, 148.
58. Merleau-Ponty, *Institution and Passivity*, 176.
59. Merleau-Ponty, *Phenomenology of Perception*, 168.
60. Merleau-Ponty, *Institution and Passivity*, 169.

61. Ibid., 118.
62. Cf. Jessica Benjamin, "Beyond Doer and Done to: An Intersubjective View of Thirdness," *Psychoanalytic Quarterly* LXXIII (2004): 5–46.
63. This does not, however, mean, as Paul Ricoeur suggests, that the unconscious is an entirely hermeneutic phenomenon, a linguistic signifier co-constructed in the analytic process. As he writes: "[T]he unconscious is an object in the sense that it is 'constituted' by the totality of hermeneutic procedures by which it is deciphered." *The Conflict of Interpretations*, 104. If it is thus a signifier, this is embodied, and thus disclosed in interpretation rather than being constituted as such by hermeneutic procedures.
64. Merleau-Ponty, *Signs*, 241.
65. Judith Butler, *Precarious Life* (London: Verso Books, 2004), 20.

3 Torture and Traumatic Dehiscence

Améry and Fanon on Bodily Vulnerability

Alexandra Magearu

Introduction

Austrian Jewish philosopher Jean Améry was arrested by the Gestapo in 1943 for his participation in the Belgian resistance.[1] After being tortured at Fort Breendonk, and subsequently deported to Auschwitz, Améry survived the ordeals of the concentration camp and published *At the Mind's Limit* (1966),[2] his exploration of the horrors of the Holocaust and National Socialism. His testimony of the torture he suffered at the hands of the Gestapo marks an arresting intervention in the philosophy of bodily vulnerability and was inspired, in part, by the writings of Frantz Fanon.[3] Upon the completion of his doctoral thesis in France, Martinican thinker Frantz Fanon obtained a job as a psychiatrist in the hospital of Blida-Joinville in Algeria (1953–1956). His analysis of the relationship between the French metropolis and its colonies, his medical observations of the traumatic effects of colonial violence upon the psyche of colonizer and colonized, as well as his implication in the Algerian liberation struggle, led him to produce some of the most profound and influential texts of decolonization, *Black Skin White Masks* (1952) and *The Wretched of the Earth* (1961).[4]

In this chapter, I explore the complications and revisions the writings of Améry and Fanon bring to Maurice Merleau-Ponty's phenomenological accounts of embodiment and flesh and Jean-Paul Sartre's theories of the look and of colonial violence. Reading Améry's reflections alongside Fanon's writings, I propose a theoretical framework for thinking about the destabilizing phenomenological effects of violence upon the body, the profound existential displacement produced by torture, and the traces it leaves on the body, often registered as fragile inner wounds that can be re-opened upon further shocks or through the psychosomatic re-experiencing of torture. I use the term "traumatic dehiscence" to refer to these existential phenomena of torture, appropriating and recontextualizing *déhiscence*, a term Merleau-Ponty employs to describe the interfolding of the sensible and sentient properties of the body. Finally, this chapter proposes that we read Améry and Fanon's theorizations of the objectified body through racialization, torture and colonial violence, against the grain of a phenomenological model of the body synergistically connected to its environment.

By engaging in a comparative study of Améry and Fanon's phenomenological accounts of vulnerability, I do not mean to collapse the historical specificity of the systems of violence that have produced the experience of the Holocaust and that of colonialism in Algeria, in relation to which they respectively write. What I am interested in investigating are the parallels they draw between different forms of sovereign power based upon the hierarchical racial differentiation of human beings that enable a type of systematic violence conducive to dehumanization, torture, massacres and genocide. In *The Origins of Totalitarianism*, Hannah Arendt argued that the racial doctrines of Nazi Germany must be analyzed in the context of the history of European imperialism and Eurocentric conceptions of the human.[5] As some of the texts discussed in this chapter will show, torture was not only the essence of National Socialism (Améry), but also a fundamental necessity of systematic colonial violence (Fanon).

Jean Améry Reads Fanon's "The Lived Experience of the Black Man"

In 1951, Jean Améry encountered Frantz Fanon's essay, "The Lived Experience of the Black Man,"[6] and he was profoundly moved and inspired by the text because of the similarities he ascertained between the condition of the colonized, Black man in the midst of the predominantly white metropolis of Europe and his own experience as a Jewish inmate in the concentration camp at Auschwitz. In his later engagement with the Fanonian corpus, "The Birth of Man from the Spirit of Violence: Frantz Fanon the Revolutionary" (1971), Améry explores the existential and historical dimensions of Fanon's theory of counter-violence mobilized against systems of repression and racialization. What he finds compelling about Fanon's writing is the fact that his philosophy of revolutionary violence is rooted in the lived experience of a colonized subject and calls for an extensive critique of European humanism and cultural hegemony. Améry writes:

> The situation in which I found myself as a Jewish concentration camp inmate was quite comparable with that of the colonised as Fanon characterised it. I too suffered repressive violence without buffering or mitigating mediation. The world of the concentration camp too was a Manichean one: virtue was housed in the SS-blocks, profligacy, stupidity, malignance and laziness in the inmates' barracks. Our gaze onto the SS-city was one of "envy" and "lust" as well. As with the colonised Fanon, each of us fantasised at least once a day of taking the place of the oppressor.[7]

Améry is referencing here Fanon's famous description of the Manichean compartmentalization of colonial space in *The Wretched of the Earth* (1961), which could be seen, for instance, in colonial Algiers—the division between the Kasbah, or the Algerian quarter, and the European quarter was symbolic of the economic and institutional segregation set in place by a colonial system based on racial, cultural and religious discrimination. In Fanon's polemical analyses, colonial Manicheanism transcends spatiality and permeates the existential realm through the demonizing of the colonized: "the colonist turns the colonized into a kind of quintessence of evil [. . .] he is reduced to the state of an animal."[8] The psychic and affective contradictions of the civilizing mission of European colonialism, which celebrates universal humanism yet dehumanizes the colonized, are met by the growing revolutionary movement of decolonization. The force of the anti-colonial struggle is born out of the violence internalized and registered by the body, and returns as violence in the form of release and resistance: "the colonized's affectivity is kept on edge like a running sore flinching from a caustic agent. And the psyche reacts, is obliterated, and finds an outlet through muscular spasms that have caused many an expert to classify the colonized as hysterical."[9] Paraphrasing Jean-Paul Sartre's formulation in the Preface to *The Wretched of the Earth*, this is the nervous condition of the colonized, which manifests itself as an embodied surfacing of repressed violence.[10]

Frantz Fanon's description of the lived experience of racialization in *Black Skin, White Masks* (1952) prefaces the analysis of the psycho-affective traumatic effects of colonial violence in *The Wretched of the Earth*. Navigating the space of the colonial metropolis for the first time, Fanon discovers that to be fixed by the white gaze is to be turned into an object amongst other objects. Fanon demonstrates that some classical phenomenological assumptions privilege a model of the human subject presumed to have an undisputed sense of the world, to be in control of one's own existence, to have an unassailable access to freedom, to be in alignment with one's own body, and to be clearly oriented in the world. Here Fanon is both inspired by and indirectly revising a number of key phenomenological insights by Maurice Merleau-Ponty and Jean-Paul Sartre.

Merleau-Ponty's *Phenomenology of Perception* (1945) proposes a model of thinking about the body as both intertwined with consciousness and as the threshold through which perceptions about the world are experienced and relations to objects are established. For Merleau-Ponty, phenomenology entails a study of the field of perception that takes into account the particularity of the embeddedness of the perceiving body, the extent of its perceptual specificity and its orientation in space and in relation to objects. Crucially, for Merleau-Ponty, a body is not an object. It is distinguishable from objects through its capacity to gaze at and over objects and, thereby, establish its perspectival horizon.[11] Because the body is able to have a perspective upon the world and because it is the fundamental vehicle for being in the world, it is simultaneously immanent and intentional. In other words, it is embedded in its milieu in specific, differentiated ways and it is the pivotal point through which objects are perceived.

The separation between the body and its perceived objects is key to Merleau-Ponty's theorization of orientation and the spatiality of one's own body. Bodily space differentiates itself from external space by virtue of being a vantage point characterized by the lived unity of the senses and motor functions of the body. Here, Merleau-Ponty deploys a concept popularized by classical psychology and neurology, the "body schema" (*schéma corporel*), and redefines it in the process as "an in principle unity" of the body, "the global awareness of my posture in the inter-sensory world"[12] and "a manner of expressing that my body is in and toward the world."[13] The body schema encompasses one's awareness of one's body in a holistic sense and ensures its situational spatiality: the relationship between different limbs, their movement and their position in relation to other objects form a general synthesis of the body and solidify the body in its distinctiveness from the objective world. Having an implicit sense of one's own body schema means being oriented toward the world, being able to express one's own intentionality toward objects and toward one's own milieu and, consequently, being able to develop impressions and observations about the world. The body schema is reworked and reinforced through the manner in which the body inhab-its space, or the manner in which it habituates itself to the world through its continuous embeddedness in the world. This leads to an understand-ing of the body as an accumulation of history, whereby the history of the body is defined as the sedimentation of bodily habits in relation to a suc-cession of experiences in the world. While Merleau-Ponty acknowledges that the experience of the body schema is differentiated in so far as differ-ent exceptional bodies are concerned,[14] his theorization of the relation-ship between thought, embodiment and lived experience is grounded in the assumption that the average able body experiences itself as a unified organism. Merleau-Ponty's insights with regard to the particularity and historicity of lived experience are crucial to understanding processes of racialization, but his account of the body schema as an in principle unity betrays the residues of universalization, and can be reworked produc-tively through the questions philosophers of race such as Frantz Fanon have raised.

Undifferentiated conceptions of lived experience lack an analysis of the socio-historical regimes of power that exclude marginalized groups from normative models of the human, limit the movement of their bod-ies in space and engender their alienation. "In the white world," Fanon writes, "the man of color encounters difficulties in elaborating his body schema. The image of one's body is solely negating. It's an image in the third person. All around the body reigns an atmosphere of certain uncertainty."[15] The moment of racialization breaks down the structure of the bodily schema and reveals a different foundational schema, what Fanon calls a "historical-racial schema" [*un schéma historico-racial*], which endows the body of the black man with negative meanings, thereby circumscribing what a body can do within the discursive and representational constraints of the white mythos.[16] The overwhelming burden of stereotypical representations of blackness disrupts the smooth functioning of affect, the tactile, kinaesthetic and visual sensations of the body, and paralyzes the senses, thereby disorienting the body by impeding the solidification of its habits and the alignment of its motor functions. This takes place through the displacement of a sense of bodily integrity (the implicit unity of the body schema) by the discursive forms of representation of the colonial imaginary, embodied in intersubjective affective encounters or materialized in the discriminatory function of colonial institutions and colonial space.

Fanon relies on Jean-Paul Sartre's ontological description of the experience of *being-for-others*, which entails encountering the looks of others and being objectified by their gaze, therefore, becoming alienated from oneself.[17] Sartre considers the look of the other (*le regard*) fundamentally unsettling because it engenders shame in the self, the shame of being looked at and valued by another, it decentralizes the experience of a unified and transparent self and it returns the subject to him/herself as an object to his/her own consciousness. According to Sartre, we have no control over the process of objectification, in other words, over the value qualifications imposed by the other upon our bodies. In this sense, because the self can become merely an instrument for the means of the other, the act of being looked at is fundamentally dangerous because it deprives the self of freedom.[18] Fanon complicates Sartre's account by arguing that the experience of objectification is highly differentiated for the Black subject because it relies upon racial taxonomies and racist representations that already haunt the differentiation between self and other. In fact, the process of racialization played out through the function of the white gaze and the white imaginary is always already over-determined by a postulation of the European subject as self-identical through his differentiation from the colonized subject. Fanon, thus, outlines a contextual and historically specific articulation of objectification—the process of racialized objectification that reproduces the alienation and marginalization of black bodies through the stabilization and empowerment of white bodies. In this sense, depending on our position in relation to the structures of racial, national, religious or gendered oppression, our sense of embodied self can be either reinforced or diminished by the gaze of the other. Key to Fanon's intervention in phenomenological debates is the insight that the lived experience of *being-for-others* can take place within a field of power structured by racial constructions, in which some bodies are devalued, marginalized and abused, while others are bolstered, confirmed and reproduced by the social milieu.[19] What makes objectification psychologically overwhelming in the process of racialization is the sense of inhabiting a "white world," in other words, being implicated in a social spatiality and a system of meanings over which one has no control, but which shapes, manipulates and imprisons the racialized body, as Fanon writes:

> As a result, the body schema, attacked in several places, collapsed, giving way to an epidermal racial schema [*schéma épidermique racial*]. . . . Disoriented, incapable of confronting the Other, the white man, who had no scruples about imprisoning me, I transported myself on that particular day far, very far from myself, and gave myself up as an object.[20]

Not only does Fanon's alter-ego become an object for the other, but he begins to experience himself as an object in the world through the processes of self-objectification in which a manufactured sense of inferiority is internalized as the origin of the self, what he describes as the epidermalization of inferiority. The internalization of objectification transpires through the skin, in other words, it manifests in the shift of the functions of the body, its capacity to move in space and its extension in the world. Cultural objectification and self-objectification are co-constitutive and continuous processes that diminish, restrict and negate the subject and her expression of agency, as well as her creativity. In the following section, I will show that Améry's account of torture cannot be fully grasped without this prior understanding of the existential and affective effects of objectification and self-negation.

Améry's Phenomenological Account of Torture

In his essay on torture, Améry writes that the first blow received at the police station is registered as a destabilizing introduction to the powers of state violence. The person who is beaten experiences a fundamental existential displacement with the first blow, what Améry calls the loss of "trust in the world," the breaking of a social contract that promises the other person will respect your bodily integrity and autonomy and care for you: "My skin surface shields me against the external world. If I am to have trust, I must feel on it only what I *want* to feel."[21] The fundamental trust that the other will have your well-being in mind is negated in the first display of violence in which your body is rendered utterly vulnerable. The beating is experienced almost as a form of sexual violence, in which the other imposes their corporeality upon the body of the victim.[22] There is not only a break in the trust invested in the other person, but also a splitting in the surface that delimits and separates the body of the person experiencing violence and the rest of the world. Skin becomes permeable, unraveled, used against the body itself to inflict pain. The invisible borders of the body and its autonomy are challenged as the space available for the body to extend and orientate itself in the world is taken over, colonized by an invasive and unwelcome presence. The symbolic moment of the first blow represents a form of rupture after which everything becomes possible because, as Améry notes, the expectation of help in the situation of bodily injury, an experience so fundamental to the constitution of the human being, is denied to the person subjected to violence.

Arrested for distributing anti-Nazi fliers to German soldiers, Améry knows nothing of substance about the workings of the Belgian resistance and is, thus, taken further to Fort Breendonk where he is tortured by the SS by being suspended from the ceiling. His arms, which had been tied behind his back, become dislocated as his shoulder joints crackle and splinter under his body weight. Because the exact feeling of physical pain cannot be faithfully conveyed in language, Améry details instead the phenomenological effects of torture:

> Whoever is overcome by pain through torture experiences his body as never before. In self-negation, his flesh becomes a total reality. Partially, torture is one of those life experiences that in a milder form present themselves also to the consciousness of the patient who is awaiting help, and the popular saying according to which we feel well as long as we do not feel our body does indeed express an undeniable truth. But only in torture does the transformation of the person into flesh become complete. Frail in the face of violence, yelling out in pain, awaiting no help, capable of no resistance, the tortured person is only a body, and nothing else beside that.[23]

Torture resembles the violence of beating at the police station in that it encompasses the violation of bodily borders by the other without any means for the person tortured to seek help or to resist the violence, but it is also an intensification of bodily vulnerability. It remolds the body beyond self-recognition and leaves indelible marks behind even when the body does not retain any visible scars. "Whoever was tortured, stays tortured," adds Améry.[24] The psychosomatic and post-traumatic effects of torture endure in the body. This is because torture brings the subject face to face with his/her own vulnerability in a gesture that concomitantly allows the other to turn him/her into mere body and forces him/her to experience him/herself on the very limits of dissolution, experiencing death while still alive, in Améry's formulation. Torture leaves the subject at the mercy of the sovereign power of the other. Moreover, because sovereign power is predicated upon deciding whether one will live or die, it reproduces itself through the transformation of the body of the victim into flesh by denying him/her possibility of transcendence in the staging of torture.

How are we to understand the specificity of torture in a phenomenological sense considering Améry's striking claim that torture consists of *transforming the person into flesh*? In more recent phenomenological studies, the experience of pain is described as an intensification of the presence of the body. Drew Leder, for instance, argues that the general experience we have of our own bodies on an everyday basis is that of absence or disappearance when it functions unproblematically, whereas it tends to be foregrounded when it malfunctions in the case of pain or disease: "pain effects what I will term *sensory intensification*."[25] James Aho and Kevin Aho claim that "pain reminds us of our own mortality"[26] and enables us to feel our bodies as paradoxically both alien and intimate at the same time—this pain which is like an alien presence in my body is also my own.[27] For Sara Ahmed, the borders of the skin become produced, fixed and impressed upon through the experience of pain:

> I become aware of my body as having a surface only in the event of feeling discomfort (prickly sensations, cramps) that becomes transformed into pain through an act of reading and recognition ("it hurts!"), which is also a judgement ("it is bad!"). The recognition of a sensation as being painful (from "it hurts" to "it is bad" to "move away") also involves the reconstitution of bodily space, as the reorientation of the bodily relation to that which gets attributed as the cause of pain.[28]

In the act of extreme violence represented by torture, which does not simply account for a pain that can be described as physical discomfort, but represents a complex affective displacement of the subject, the borders of the body become vividly real as they are not only used to inflict pain but also transgressed by the torturer. Because the attention, intentionality and orientation of the body become intensely focused on the source of pain impressed upon the body, Améry suggests that consciousness is on the verge of dissolving into mere flesh.

Elisabeth Weber's discussion of Améry's German terminology is very resourceful here: "One of the words used by Améry stands out: the word rendered in the English translation as 'transformation of the person into flesh.' Other than in theological debates about the incarnation of Christ, it is a word rarely used in German: *Verfleischlichung*."[29] Moreover, Weber notes that the word, which means indeed transformation into flesh, is redefined by Améry as "an intensified incarnation—but an incarnation out of which any spirit has been driven."[30] Weber prefers to translate the term as "fleshization" or "becoming meat" in order to reference Deleuze's discussion of meat as the common zone between human and animal[31] and, thus, to note the dehumanization enacted by torture.[32] Building on

Weber's argument, I would like to propose reading *Verfleischlichung* as a tension between enfleshment and objectification. While the English term "enfleshment" carries the theological ramifications of the incarnation of Christ (similarly to the German *Verfleischlichung*) and notes the martyrdom produced by the impunity of sovereign violence, *objectification* encapsulates the stripping of the body to bare life through the extreme violence of torture, its transformation into an object amongst objects, and its denial of spiritual life. In this sense, Améry's understanding of the body objectified by extreme violence can be analyzed in the context of Sartre's discussion of objectification as a denial of the transcendent dimensions of the body, which should be coupled with a Fanonian reading of the racializing effects of nation state violence, as I argued earlier. Reading this episode alongside Améry's Fanonian reflections on the space established by the Nazi concentration camp as a materialization of racial violence, torture emerges as one of the main tools of sovereign power through which subjects are dehumanized.

Furthermore, *Verfleischlichung* recalls, and arguably recontextualizes, Maurice Merleau-Ponty's term, *chair du monde* (the flesh of the world), developed as part of his later ontology of flesh. In his influential essay, "The Intertwining-The Chiasm," published posthumously as part of *The Visible and the Invisible* (1968),[33] Merleau-Ponty introduces his conception of flesh as the elemental medium through which the body interacts with the world. The French philosopher now describes the body as doublefold in its constitution, a two-dimensional being that is said to be part of the sensible world as an objective body amongst other bodies *and* to be a sensible body capable of seeing and touching. The example he employs to demonstrate the intricate constitution of the body is the event in which our right hand touches the left and we can both feel that we are touching and that we are touched at the same time. There is a reversibility to touch as there is to our own bodily surface.[34] In search for means of avoiding dualism, Merleau-Ponty finally comes to the following formulation:

> If one wants metaphors, it would be better to say that the body sensed and the body sentient are as the obverse and the reverse, or again, as two segments of one sole circular course which goes above from left to right and below from right to left, but which is but one sole movement in its two phases.[35]

The sensible and sentient properties of the body are not discrete and autonomous processes, but they form a circular movement of continuous feedback, which entails the co-constitutive makeup of the body. In other words, the perceptual knowledge derived from the sentient body is informed and shaped by the experience of the body sensed, and conversely, the specificity of the body sensed determines the manner in which perceptual inputs are registered by the sentient body. Elizabeth Grosz's appropriation of the image of *the Möbius strip* from Jacques Lacan topological work seems most suitable here to illustrate Merleau-Ponty's model of the body.[36] If you follow the surface of the Möbius strip with your fingers the inside turns into outside and the outside turns back into inside. The body (and its visible and tangible properties) functions similarly to the Möbius strip in that its perceptual input folds into and forms the perceiving senses and vice versa in the manner of a feedback loop.

The flesh, for Merleau-Ponty, is what makes the reversibility of the body possible. It is not a fact of matter, neither a substance, nor a product of the mind. Flesh is instead "an element of Being" which mediates between the visible and the invisible and reveals the complex constitution of the body:

> the flesh we are speaking of is not matter. It is the coiling over of the visible upon the seeing body, of the tangible upon the touching body, which is attested in particular when the body sees itself, touches itself seeing and touching the things, such that, simultaneously, *as* tangible it descends among them, *as* touching it dominates them all and draws this relationship and even this double relationship from itself, by dehiscence or fission of its own mass.[37]

At the intertwining of the sentient body and the sensed body, flesh denotes a relationship of similarity and difference with the world of objects, in that contact is possible because the body can simultaneously be an object amongst other objects, while remaining an ordering consciousness. Flesh is an ultimate notion for Merleau-Ponty, in that it exists in the form of a flow that traverses the body and the world. The flesh of the world is an expression that describes the synergetic sensual relationship between bodies in the world, whose landscapes interweave.

"Dehiscence" is an intriguing term derived from botany, which refers to the phenomenon through which certain plants' pods burst open in specific places in order to discharge seeds, pollen or fruit. For Merleau-Ponty, dehiscence (French: *déhiscence*) signifies the splitting open of the body in two, for instance in the mutually constitutive experience of touching and being touched, so that there is no centralized internal focus of perception but an encroachment or a crisscrossing of the perceiver and perceived.[38] The term derives from the Modern Latin *dehiscentia* and *dehiscentem*, the present participle of *dehiscere* which translates as "to gape, open or split down." It is also used in medical parlance to refer to *wound dehiscence*, the reopening of formerly sutured wounds. It is to this latter meaning, which haunts Merleau-Ponty's ontology of flesh, I would now like to turn as a way of analyzing the place of experiences of pain and violence in the phenomenology of embodiment.

The harmonious extension of the self into the world as flesh as part of intersubjectivity is only possible if there is a reciprocity between self and other in the mutually constituted relationships of seeing and being seen, touching and being touched, hearing and being heard. In becoming flesh through experiences of violence, the subject is pressed to the outer layers of their being, coerced to inhabit the borders of their body as they become reshaped by the experience of being confined, invaded and harmed by the other. The permeability of bodily borders, fundamental to the participation of the body in the flesh of the world, is now coaxed into painful materiality as an imposed boundary produced by violence. In physical torture, the body of the tortured is severed from her relationality to the world through the intensification of affect upon the surface of the skin where pain is impressed. Or, if the body's position is used to inflict internal harm to the body, as is the case with Améry's suspension that makes his shoulder joints give way under the weight of his body, the orientation of the body is all concentrated in the feeling of dislocation. There is a *traumatic dehiscence* that takes place, then, with the opening of external and internal wounds in the body, in which the dialectical relationship between the sensible and sentient properties of the body malfunctions. The transformation of the person into flesh encompasses the complex objectification of the body enacted by torture in which the sentient capacities of the body are forcefully channeled toward the intensified experience of pain, therefore foreclosing other relational connections the body normally establishes with the world, as well as the creative, transcendental and agential properties of the body. This fundamental existential displacement of torture can remain lingering in the body through the post-traumatic effects of the original violence of severance.

Sartre and Fanon on Torture in the Algerian War of Independence

In his introduction to Henri Alleg's testimony of surviving torture by the French army in Algeria,[39] Jean-Paul Sartre deplored the metropolitan French population's silence and complicity in the horrors their government inflicted during the Algerian War, a protracted conflict which lasted from 1954 until the independence of Algeria in 1962. The unwillingness to take responsibility for state violence was blatant to him because, as he noted, not long before the French had been victims of torture themselves in the hands of the Gestapo occupation forces in Paris. The analogy between the violence of the Third Reich and French colonialism in Algeria was often raised by left-leaning French intellectuals of the time as a means of opposing what was known as "the gangrene" of French society, the use of torture with impunity. Fanon, writing in *El Moudjahid*,[40] expressed his disenchantment with these displays of solidarity for the sake and in the name of salvaging "French honor," which appeared often ineffective to him because of the emphasis placed on the French sensibility and the erasure of the actual victims of the extensive military torture complex, Algerian men and women.[41] Yet Sartre, distinguished himself by providing trenchant critiques of French colonialism and its policies in Algeria in his work, and by maintaining close intellectual relationships with anti-colonial thinkers such as Frantz Fanon and Albert Memmi.

For Sartre, torture in Algeria "was simply an expression of racial hatred."[42] The dehumanization enacted by torture was a symptom of a larger discursive system which backed the claims of colonists to cultural superiority, a mystification of the economic relations of exploitation and its transmutation into a doctrine of human hierarchies. Fundamentally, for Sartre, the colonial system in Algeria was a structure of economic and existential dependency—the exploiters were dependent on their victims for their wealth, yet this relation of power was repressed only to return in the form of racism. The emancipation of the colonized threatened not only the financial security of colonialists, but also their exclusive claim to humanity.[43] Moreover, as he notes in his later reflections on colonial violence in the *Critique of Dialectical Reason* (1960), colonialism enacts a double alienation, not only that of the colonized who are estranged from themselves through the contradiction of forced assimilation and discrimination, but also of the colonists themselves who quench their repressed anxieties about losing their domination foothold through racism as praxis:

> the colonialist reveals the violence of the native, even in his passivity, as the obvious consequence of his own violence and as its sole justification. This discovery is made through hatred and fear, as a negative determination of the practical field, as a co-efficient of adversity affecting certain multiplicities in this field, in short, as a permanent danger which has to be avoided or prevented. Racism has to become a practice: it is not contemplation awakening the significations engraved on things; it is in itself self-justifying violence: violence presenting itself as induced violence, counter-violence and legitimate defence.[44]

In this cycle of violence, the institutionalization of racism through torture becomes a form of counter-insurgency. The justification of torture often invoked as legitimate defense by the French army—that of saving countless lives from terrorism through the torturing of one person—obfuscates the function of torture as a means of suppressing the expressions of resistance of the colonized, and, indeed, as a method of disciplining Algerians into passive participation in the workings of colonialism.

Torture was not an exception, but the rule of the French strategy in Algeria. For Frantz Fanon, torture was a fundamental necessity of the colonial world, "an expression and a means of the occupant-occupied relationship."[45] Fanon, who was not tortured himself, closely observed the psychosomatic and traumatic effects of torture in his patients at

the Psychiatric Hospital of Blida-Joinville and at the National Liberation Army's medical facilities. In comments resembling Sartre's allusions
to the double alienation of colonialism, Fanon writes about the effects
of the violence of the war on both victims of torture or survivors of
massacres *and* on police officers responsible for torture. A 37-year-old
Algerian peasant, who survives a massacre of the French army in his
village, develops random homicidal impulses and a delirious desire to
"kill everybody," including the ALN (Armée de Libération Nationale)
soldiers who had taken him under their protection. A former student and
ALN fighter suffers from severe depression and depersonalization due to
being haunted by the image of a French woman he killed in a displaced
moment of retribution for the killing of his mother at point-blank range
by a French soldier. A 20-year-old European police officer is referred to
the hospital by his superiors because he presents behavioral problems.
The man suffers from depression and he is troubled because he can continuously hear the screaming of the people he tortured, especially the
ones who died in custody. In a particularly distressing scene recounted by
Fanon, the police officer encounters one of his former victims who was
treated for post-traumatic stress at the same hospital—while the French
policeman is seized by a panic attack, the Algerian man is later discovered hiding in the bathroom and trying to commit suicide in order to
avoid being captured again.[46] Finally, a French police inspector displaces
the violence internalized through his day-long torture sessions at work
onto beating and torturing his wife and children at home. These cases
demonstrate that the violence of torture transcends the intersubjective
relationship between torturer and tortured. Torture remains lodged in
the body in the form of auditory hallucinations, anxiety, post-traumatic
stress and depression. Moreover, it permeates all levels of life in its multiple displacements and reconfigurations, affecting the extended social
circles of torturers and tortured.

Fanon also records the psychological, affective and embodied symptoms experienced by his patients who have undergone torture. The
experience of torture returns indeed like the reopening of the wound in
dehiscence through the reenactment of the physical sensations impressed
upon the body of the tortured, as well as through heightened emotional
states associated with trauma. After torture with electricity, one of the
most routinely employed methods in French prisons in Algeria, patients
experience "local or systemic somatic delusions" in which they "feel pins
and needles throughout the body and get the impression their hands are
being torn off, their heads are bursting, and they are swallowing their
tongue."[47] In addition to generalized apathy and lack of energy, they also
develop a "phobia of electricity" in which they fear touching the light
switch, radio or telephone. After the administration of the drug known
as Pentothal, a chemical used as a truth serum during interrogation and
torture sessions, patients present repetitive verbal ticks, blurred mental

and sensory perception manifested in their inability to recognize the existence of objects or tell the difference between truth and falsehood, inhibition and a phobia for one-to-one conversations. Noting the persevering effects of other psychosomatic disorders such as stomach ulcers, disturbed menstrual cycles and premature hair whitening, Fanon writes that they are evidence of the fact "that there is no need to be wounded by a bullet to suffer from the effects of war in the body and soul."[48] The invisible scars of war are retained by the psyche, yet they often resurface in bodily manifestations that transport the sufferer back to the event of torture.

Conclusions: Traumatic Dehiscence as the After-Life of Bodily Injury

In his reflections on the neuroses produced during times of war, Sigmund Freud uses an economic model of affect to refer to the dynamic of traumatic mental processes characterized by fixation to the moment of the traumatic accident and a recurrent revisiting of the episode in dreams or in panic attacks:

> Indeed, the term "traumatic" has no other sense than an economic one. We apply it to an experience which within a short period of time presents the mind with an increase of stimulus too powerful to be dealt with or worked off in the normal way, and this must result in permanent disturbances of the manner in which the energy operates.[49]

The traumatic dehiscence enacted by torture can thus be conceived as a transfer of energy which no longer functions normally, but which breaks down, leaving indelible wounds within the psyche and upon the body. If the dialectical relationship between the interfolding sensible and sentient capacities of the body was premised upon the embeddedness of the body in the flesh of the world, as a participation in a continuous, unhindered flow of energy, the broken body is assaulted by an overflow of stimuli which reshapes and reconfigures its capacities. The reopening of the wound in the later triggering of the initial trauma acts, in Freud's words, as a "complete transplanting of the patient into the traumatic situation."[50]

In this context, Fanon's detailed observations of the post-traumatic effects of torture are invaluable for understanding the temporal dynamic of traumatic dehiscence—not only has the tortured victim experienced a traumatizing splitting, blocking some of his sensorial capacities and impressing pain deep within the body, but he experiences himself stretched back in time and confined to the event of torture through the lingering physiological and affective remnants of torture. The experience of the body in alignment with its corporeal schema malfunctions as phantom pains and phobias continue to haunt the traumatized subject.

In one of the most memorable passages of his essay on torture, Améry claims that "whoever has succumbed to torture can no longer feel at home in the world."[51] The term used by Améry in the original German text, "nicht mehr *heimisch* werden in der Welt,"[52] is reminiscent of a term Martin Heidegger employs in his philosophy of Being: the uncanny (*unheimlich*). In fact, Améry's choice of language to describe torture throughout his work invokes Heidegger's existential phenomenology of being-in-the-world, while simultaneously taking a departure from it. For Heidegger, anxiety, as the awareness of Dasein of the world, manifests as a feeling of displacement, "not-being-at-home" ("Nicht-zuhause-sein").[53] This is, according to Heidegger, a primordial phenomenon of being-in-the-world, which reveals the very fact of Dasein's being thrown in the world as its fundamental experience of the world. I infer that Améry revises Heidegger's notion of being-in-the-world in order to emphasize the specificity of the traumatic displacement produced by torture and, thus, to make visible forms of violence Heidegger ignored and remained silent about, the genocidal cruelty of National Socialism.[54] Once one has experienced torture, one remains tortured, in Améry's understanding, because the fundamental trust in the world that allowed for one's harmonious embeddedness in the world has been permanently broken. Torture, then, in its self-negation, remains in the body in the form of alienation from the world. Yet, unlike Elaine Scarry's claim that intense pain is both world-destroying and language-destroying,[55] Améry shows that, at least in his case, a recovery of voice and a reconfiguration of one's personal experience and being-in-the-world is possible through writing. His intensely personal testimony raises fundamental challenges to our conceptions of violence and modernity.

Notes

1. Born Hans Chaim Maier in a family of assimilated Jews in Vienna, Améry later changes his name to a Francophone anagram of his last name in a symbolic gesture of breaking with the German tradition. His work also reflects his unease with what he calls the "necessity and impossibility of being a Jew," his disconnectedness from Jewish religion, culture and thought coupled with his interpellation as a Jew by the Nuremberg Laws passed by the National Socialist German state.
2. Améry's original German-language text, *Jenseits von Schuld und Sühne*, was translated as *At the Mind's Limits* in 1980.
3. I am thankful to Paul Gilroy's article, "Fanon and Améry: Theory, Torture and the Prospect of Humanism," for pointing out the intersections between Fanon and Améry's work.
4. Fanon's texts were originally published in French as *Peau noire, masques blancs* and *Les Damnés de la terre*, respectively.
5. Hannah Arendt, *The Origins of Totalitarianism* (New York: Harcourt Books, 1966 [1951]), 158.

64   *Alexandra Magearu*

6. The essay was circulated in 1951 and was later published as part of Fanon's work, *Black Skin, White Mask* (1952).
7. Jean Améry, "The Birth of Man from the Spirit of Violence: Frantz Fanon the Revolutionary," trans. Adrian Daub, *Wasafiri* 20, no. 44 (2005 [1971]): 15–16.
8. Frantz Fanon, *The Wretched of the Earth*, trans. Richard Philcox (New York: Groove Press, 2008 [1961]), 6–7.
9. Ibid., 19.
10. Jean-Paul Sartre, *Preface to The Wretched of the Earth*, trans. Richard Philcox (New York: Groove Press, 2008 [1961]), liv.
11. Maurice Merleau-Ponty, *Phenomenology of Perception*, trans. Donald A. Landes (New York: Routledge, 2012 [1945]), 96.
12. Ibid., 102.
13. Ibid., 103.
14. See for instance Merleau-Ponty's observations about atypical bodies such as those that experience the phenomenon of the phantom limb (81).
15. Frantz Fanon, *Black Skin, White Masks*, trans. Richard Philcox (New York: Grove Press, 2008 [1952]), 90.
16. Ibid., 91.
17. Jean-Paul Sartre, *Being and Nothingness*, trans. Hazel E. Barnes (New York: Washington Square Press, 1992 [1943]), 345.
18. Ibid., 385.
19. For a further exploration of these issues, see Sara Ahmed's phenomenological discussion of spatiality and difference in *Queer Phenomenologies: Orientations, Objects, Others* (Durham: Duke University Press, 2006).
20. Fanon, *Black Skin, White Masks*, 92.
21. Jean Améry, *At the Mind's Limits*, trans. Sidney Rosenfeld and Stella P. Rosenfeld (Bloomigton: Indiana University Press, 1980 [1966]), 28.
22. Arguably, Améry's conflation of the violence of beating and that experienced in rape erases the gendered specificity of sexualized violence as a form of torture. Gisèle Halimi and Simone de Beauvoir, for instance, emphasize the specific dimensions of gendered torture in *Djamila Boupacha*, a report they wrote in support of the trial of Djamila, an Algerian woman who was captured by the French forces during the Algerian war, tortured and raped with a bottle (1962).
23. Ibid., 33.
24. Ibid., 34.
25. Drew Leder, *The Absent Body* (Chicago: Chicago University Press, 1990), 71.
26. James Aho and Kevin Aho, *Body Matters: A Phenomenology of Sickness, Disease, and Illness* (Lanham: Lexington Books, 2008), 6.
27. Ibid., 19.
28. Sara Ahmed, *The Cultural Politics of Emotion* (New York: Routledge, 2004), 26.
29. Elisabeth Weber, *Kill Boxes: Facing the Legacy of US-Sponsored Torture, Indefinite Detention, and Drone Warfare* (New York: Punctum Books, 2017), 58.
30. Ibid., 59.
31. See Gilles Deleuze, *Francis Bacon: The Logic of Sensation*, trans. Daniel W. Smith (Minneapolis: University of Minnesota Press, 2004 [1981]).
32. Ibid., 60.
33. The French edition, *Le visible et l'invisible*, was assembled by Claude Lefort in 1964 from unfinished notes left upon Merleau-Ponty's death in 1961. The text is therefore incomplete and assembled into an edited collection posthumously.

34. Maurice Merleau-Ponty, *The Visible and the Invisible*, trans. Alphonso Lingis (Evanston, IL: Northwestern University Press, 1968 [1964]), 137.

35. Ibid., 138.

36. In her introduction to *Volatile Bodies: Toward a Corporeal Feminism*, Grosz uses a term Lacan employed to think about the interfolding of the conscious and the unconscious, the Möbius strip, which shows for her "the inflection of mind into body and body into mind" (xii).

37. Merleau-Ponty, *The Visible and the Invisible*, 146.

38. Ibid., 123.

39. During the Algerian War, Henri Alleg was the editor-in-chief of the newspaper *Alger Républicain* and a member of the Algerian Communist Party. Arrested for his support of the Algerian liberation struggle in 1957, he was tortured by French paratroopers through repeated beatings, water-boarding, electricity and he was administered Pentothal, an experimental drug that was believed to act as a truth serum. Alleg wrote about his ordeals on snippets of paper which were eventually smuggled out of the prison and published in *La Question* (1958).

40. *El Moudjahid* is an Algerian French-language newspaper that represents the views of the Front de libération nationale (National Liberation Front), the main party involved in the Algerian liberation struggle that later became the single governing party of the postcolonial state. Fanon wrote several articles for the newspaper from 1952 to 1961, and served as its editor upon his exile in Tunisia during the war.

41. Frantz Fanon, *Toward the African Revolution*, trans. Haakon Chevalier (New York: Grove Press, 1967 [1964]), 71.

42. Jean-Paul Sartre, *Introduction to The Question* (New York: George Braziller, 1958), 33.

43. Ibid., 32.

44. Jean-Paul Sartre, *Critique of Dialectical Reason*, trans. Alan Sheridan-Smith (New York: Verso Books, 2004 [1960]), 720.

45. Fanon, *Toward the African Revolution*, 66.

46. Fanon, *The Wretched of the Earth*, 196.

47. Ibid., 211.

48. Ibid., 217.

49. Sigmund Freud, *Introductory Lectures on Psycho-Analysis*, trans. James Strachey (New York: W. W. Norton and Company, 1989 [1917]), 340–1.

50. Ibid., 341.

51. Améry, *At the Mind's Limits*, 40.

52. Jean Améry, *Jenseits von Schuld un Sühne* (Stuttgart: Klett-Cotta, 2002 [1966]), 85, my italics.

53. Martin Heidegger, *Being and Time*, trans. John Macquarrie and Edward Robinson (New York: Harper Perennial, 1962 [1927]), 233.

54. For a helpful study of Heidegger's ambivalent relationship with National Socialism, his lack of responsibility towards the Holocaust, and his anti-Semitic views revealed in his *Black Notebooks*, see Andrew Cooper's, "Beyond Heidegger: From Ontology to Action," *Thesis Eleven* 40, no. 1 (2017): 90–105. Besides the fact that Heidegger excludes Jews from Dasein in some of his antisemitic writings, his universalizing descriptions of "not-being-at-home" in the world render most experiences of alienation comparable in that they are supposedly produced by Dasein's being thrown-in-the-world. Améry's reading of the *unheimlich*, thus, draws attention to the historically specific features of genocidal power and extreme violence as political causes for existential displacement.

55. Elaine Scarry, *The Body in Pain: The Making and the Unmaking of the World* (Oxford: Oxford University Press, 1985), 35.

4 Framing Embodiment in Violent Narratives

Cassandra Falke

Over the last two centuries the percentage of civilians affected by political violence has increased. The UN estimates that, a century ago, 90% of war casualties were military personnel. By the end of World War II, 65% of casualties were non-combatants. Now, that number has risen to 90%. Ninety percent of war casualties are non-combatants.[1] It would be nice to agree with Canadian psychologist Stephen Pinker that the "better angels of our nature" are winning and the world is becoming a more peaceful place, but a great deal of historical nuance has to be ignored in order to reach that conclusion.[2] At least within the field of political violence, what was once viewed as a major ethical breach has become a norm.[3] Astounding though they are, statistics about civilians killed in political violence barely evoke the individual lives that each digit stands for. As authors in literature, media or humanitarian organizations seek to raise awareness about this violence, they seek ways of evoking the reality of individual human lives, lives that are not just "being lost," as though somehow misplaced from their position in the world, but becoming mired in brokenness, pain and fear around the event of death. The number of stories about these victims increases with the body count. Stories then proliferate on computer screens and television sets, magazines and newspapers, fliers in the mail, recounting the great amount of pain that humans cause one another to suffer, while many readers (and most readers of this book) sit safely thinking.

Reading about other people's pain produces a swarm of ethical questions. Are there more or less ethical ways of portraying other people's pain? More or less ethical ways of reading a portrayal of pain? How does directing our attention to pain caused by violence differ from directing attention to the natural pain of disease or accidents? If we read about violence that we find ethically unconscionable, does that reading make any difference for who we are or how we live after we set the reading aside? The same destruction or mauling of a person can become a story of "collateral damage," a story of tactical failure, a story of a good woman gone, a story of a mistaken act of trust or just bad luck. It becomes the tragic story a family inherits or a story of mission accomplished. Or it fails to reach the threshold of becoming a story at all.

The stories through which truths about political violence reach us are, therefore, of great importance for mobilizing ethical action in response to violence against civilians. In this chapter, I consider one element of the ways that narratives of political violence are framed—the presentation of the body. Specifically, I examine the ways that embodiment is framed in narratives designed to encourage violence in contrast to narratives that portray violence for non-violent ends or for the sake of morally neutral contemplation. Michael Staudigl suggests that "phenomenology must seek to thematize the *conditions of possibility* of violence and relations of violence; it must focus on what is responsible for rendering such . . . violence *possible* in the first case, and must ask how what is identified can then subsequently become the subject of discourses of justification."[4] This chapter contributes to such a thematization by comparing the representation of the body in three violent narratives, two of which resist assimilation into a discourse that justifies violence.

Most speculations about authorial intentionality within literary criticism are conjectural, so my categorization of texts based on intention might appear dubious, but outside of literary circles, authors often do state their intention. Two of the narratives I discuss are embedded in texts designed to persuade readers to perform a specific action relevant to political violence, so their intention is quite plain. The first text, Martin Luther King's "Letter from Birmingham Jail" articulates the reasoning behind non-violent resistance to political oppression. It was written at the height of the civil rights conflict in Birmingham, Alabama, in 1963. The second text explicitly promotes violence. It is an article about the 2016 Brussels airport bombing published in *Dabiq*, which is the magazine ISIS produces online to recruit terrorists and publicize their activity. King portrays the body as a medium through which we, as subjects, interpret ourselves in the context of everyday life, showing how that interpretive process always incorporates culturally available categories—black/white, disposable/ grievable, symbolic or lived within. In contrast, ISIS authors consistently present the body as an instrument for practical or symbolic use. The body is pictured almost exclusively in extreme moments preceding or following violence. When the unity of a narrative about an attack necessitates describing the decision to kill or decisions about who to kill, glimpses of everyday life may be included, but only for the sake of sorting perpetrators and victims into the rigid roles of martyr, mujāhidīn or infidel.

The third and final text I discuss is Joseph Conrad's *The Secret Agent*. Published in 1907, *The Secret Agent* looks back on the upsurge of anarchist activity in London in the 1880s and 1890s. It was inspired by Conrad's contemplation of a "blood-stained inanity" that had occurred at Greenwich Observatory in 1894. In an attempt to bomb the observatory, French anarchist Martial Bourdin unleashed the explosion prematurely and blew himself up with no damage at all to the intended target. Recalling this event 13 years later, Conrad produces a novel satirizing the anarchist movement and the intelligence community's response to it. Conrad's portrayal of the body in the text is notable for a couple of reasons. First, the materiality of the body is stressed throughout, both through the preponderance of physical description generally and also through the detailed description of the young man's physical remains. The "accumulation of raw material for a cannibal feast," which had been his body is described as part of the police investigation on page 64 and re-invoked several times over the succeeding 163 pages. Also, the book fore-grounds characters' very different methods of interpreting others' bodies. Whereas the young man who is killed feels an incapacitating "tenderness to all pain and misery," animal or human, one of the anarchists sees only criminologically defined physical types—"the degenerate type," "the murdering type," "the lying type."[5] Conrad's implied concern for the body as a key medium in readers' and characters' ongoing hermeneutic processes makes the novel particularly interesting for an investigation of the body in violent narratives.

Three linked, phenomenologically relevant issues emerge from this brief summary of these texts. First, there is the presentation of the body itself. Is the capacity of feeling elaborated? Is the body of everyday life connected to the body subjected to violence? To what extent is the vulner-ability of the body presented through physical descriptions of wounded or broken bodies? Second, there is the question of how categories availa-ble for interpreting the body are presented as operating in reference to the subject's understanding of him/herself, the reader's understanding of that subject, and the perpetrator's understanding of his/her victim. The extent to which perpetrator and victim are presented as themselves subjects is part of this representational process. Finally, there is the potential of con-trasting possible interpretive processes as they relate to the body. After an overview of the phenomenological grounding on which my inquiry is built, I address each of these issues with regard to each of my three texts.

Phenomenological Grounding: Other People's Bodies

Some scholars have recognized that phenomenology provides new ways of thinking about the narrativization of violence against civilians. Per-haps most famously, Judith Butler's *Precarious Lives* draws on Levinas's concept of the face to argue that "dominant forms of representation can and must be disrupted" in order for "the precariousness of life to be apprehended."[6] Her later work *Frames of War* again evokes "the face" to examine representational "frames that foreclose responsiveness."[7] Although she does not refer to Michel Henry's phenomenology of life, she echoes Henry's phenomenological search for "the trace of what is human that confounds the norm of the human," the irreplaceable and vulnerable something that is snuffed out forever when a human life is destroyed.[8] Also thinking of Levinas, Zygmunt Bauman writes in his 2011 *Collateral*

Damage: Social Inequalities in a Global Age that "casting others as 'security problems' leads to an effacing of . . . those aspects of the Other that put us in a condition of ethical responsibility."[9] His earlier *Postmodern Ethics* (1993) elaborated more generally the ways that "the *impossibility* of not being responsible for this Other" (53) initiates an ethics beyond any "standard of reciprocity, of equal treatment, of balanced exchange."[10] Butler and Bauman both recognize that phenomenology offers new ways of approaching the problem of violence because it sets aside the law-derived categories that foreclose certain forms of inquiry about violence. For instance, an inquiry that operates using the categories of victim, perpetrator and witness often begins by associating passivity with the victim, agency with the perpetrator and true recollection with the witness. One can then *discover* a victim's agency, a perpetrator's suffering or responsiveness (associated with passivity) or a witness's biased interpretation. But such a discovery only becomes possible because the terms of the inquiry had artificially defined the inseparable and dynamic push and pull of passivity/activity or experience/interpretation as separable. To quote Staudigl again, phenomenology offers a way "to think violence as thoroughly relational."[11] However isolating the experience of pain may be, as a referent in a violent narrative, it presumes intersubjectivity on two levels: the reader/text/author relationship that makes the pain of another body imaginable and the already-togetherness of all life, in which our dependence on each other precedes the illusion of independence that makes violence possible.

Phenomenology presents the body as neither an object nor a tool for transparently mediating the outer world to subjectivity. Rather, the body is that most primary means of knowing, so foundational as a way of knowing that we most often become aware of it through the processes of coming to know something else—cold water down one's back, an embrace, pain. These things arrive either from the outside of our bodies or at least from outside the experience of our bodies that we call normal. Pain arrives as a temporally bound experience that is not intrinsic to our awareness of our body because it is not always the same or even always present. Coming from elsewhere, it maps the nerves that radiate through arms and hips. But once present as part of bodily experience, pain, like other bodily experiences that enforce our awareness of a before and after, alters other bodily experiences that we might have presumed were unrelated, such as movement, wakefulness, the availability of attention for other things. Blindness awakens us not only to the sight we have lost but to sounds—the sound of blood pulsing, the constant electric background of modern life. Forces from outside of our bodies can similarly awaken bodily awareness even if they are not primarily experiences of our own bodies. While bandaging my child's cut, for example, I feel my own heart constrict. I only know the cold someone else feels because I have felt it myself.

Acts of consciousness are then born from and within experiences of the body and constantly have the body as a first language. As Merleau-Ponty puts it, "I am conscious of my body via the world" and conversely "I am conscious of the world through the medium of my body."[12] The work of contemporary phenomenologist Jean-Luc Marion complements Merleau-Ponty's ideas here. Marion asserts that I am conscious of my ego via the flesh and conversely, I am conscious of my flesh through the medium of my ego.[13] The metacognitive reflection from which all phenomenology (sometimes in spite of itself) begins redeploys what we know of our bodies and the world as means of knowing and speaking. Our bodies and the world frame and reframe our bodies and the world as we move toward self-awareness.

The most important distinction between the ways that the flesh and the world mediate consciousness, at least for the consideration of violence, is that we have, in the present, more willful possibilities with regard to our body, possibilities for embodied action, embodied feeling, and the possibility of imagining future action via our bodies. We cannot imagine or perform actions via the world with the same degree of authority. Walking in the hills, I do not expect the tree to move. I move. The tree cannot and will not yield to me. Even if I cut it down, there is no yielding. This changes, of course, when we begin to conceive the world as a shared world, a world in which the mediation of our own flesh here, the tree out there, or the concept of our own minds is always already mediated by the presence of other people—people who have taught us our body's capability or not, who have helped us become secure in touching other bodies or not, who have taught us to see the body as a tool, a medium, a gift and/or a site of self-display and self-becoming. And as Levinas and Marion would both teach us, there is never a time before the time that my becoming relied on another person. Because of the absolutely essential role other people play in the mediation of the world by our bodies and the mediation of our bodies by the world, we cannot say that we live in our flesh independently of other people. Even before we begin to think about the other people who grow the food or knit the sweaters that enable the maintenance of our bodies by manipulating other things in the world, we have to admit that we depend ontologically and epistemologically on other people for our own enfleshed self. We are born from other bodies. We interpret our own bodies as colorful, functional, beautiful or flawed in relation to other bodies through direct comparison or through language sprung from other hands and mouths.

Both Levinas and Marion emphasize that relationality precedes self-hood. For both of these philosophers, there are no clothes, no titles, no threats or scars that conceal the naked face of the other. Every time we encounter another person and even before that encounter, the mortality of the other reaches us as something fragile that we have a duty to protect.[14] There is not space here for an extended look at what makes

Marion and Levinas's formulations of embodied selfhood different, but there is one difference that affects the way we approach the problem of violence. For Levinas, relationality with the other is experienced as a form of violence at such a primary level that coming into being is itself an act of being violently exposed. He uses the language of violent victimization to discuss the relationship of the self to time: "Incarnation. . . . It is the correlate of a persecution."[15] He also uses it to describe the already-present relationship we have to others before we become conscious of ourselves. He writes in *Otherwise than Being* that "in the exposure to wounds and outrages, in the feeling proper to responsibility, the oneself is provoked as irreplaceable, as devoted to the others, without being able to resign, and thus as incarnated in order to offer itself, to suffer and to give."[16] Individuality, or the irreplaceability of the self, arises from the situatedness of that self in a position of physical responsibility for the other. That position is not *like* but *is* a position of woundedness. Because it is never enough. We may be strong enough or quick enough to interrupt or ameliorate another's pain, but never to stop it.

Marion describes the same already-being-connected. He writes that "my very own flesh (it makes me become myself, which I was unaware of before then) comes upon me . . . in the measure in which the flesh of the other provokes it. . . . This erotic inadequacy allows me to accede to myself . . . by letting myself be (dis)possessed."[17] The responsibility of giving the other to him/herself is not a decision, but, as with Levinas, something that we are born into and continually actualize consciously or not. For Marion, however, the inadequacy of a would-be autonomous self is not necessarily experienced as a state of woundedness, but may be a recognition of gifts already given, a recognition of the extent to which our continual presence at all, and certainly our continual presence as a functioning and loved person, has already been secured by others. When contrasting the language that the two use to describe our interconnectedness, it is almost impossible not to think of Levinas's personal life—his captivity in a prisoner of war camp, the death of his parents, his brothers and his wife's parents.[18] How could he not think of life in others' debt as a wound? Through that experience, he arrives at a truth about one spectrum of bodily vulnerability and consciousness, but Marion's claims about the joyous potential of that vulnerability are no less true.

There is an infinite variety of possible ways of acknowledging the responsibility given to us through shared bodily vulnerability, and an infinite number of ways to narrate each of these possible expressions. Narratives about violence always work within the awareness of bodily vulnerability (because they expose that vulnerability), but they do not always acknowledge the negation of another self or the negation of shared embodied responsibility that violent acts entail. Indeed, some stories of violence present vulnerability as something to be exploited. Among those stories that do acknowledge a connection between vulnerability and responsibility for the other, there is still variation along the spectrum of pleasure and pain associated with that acknowledgment. A narrative that awakens us to our embodied responsibility for the other may operate more in a Levinasian key, focusing readers' attention on the dangers of our vulnerability to one another. Or, narratives may direct attention to the loving acceptance of one flesh by another—the "yes" that every beloved offers the other in physical intimacy. This is the case even in violent narratives because they are often also stories of loving sacrifice or consolation after a crisis.

The complexities inherent in all of these narrative possibilities are further complicated by questions about readers' ethical responsibility toward the embodied others we read about. Fictionality cancels that responsibility, if the bodies described belong to characters in novels or poems, but what about real-life people in situations comparable to those of the characters? If reading a novel wakens insight about the plights of those real others, do readers have the same responsibility to those real others that they would to someone present in the flesh?

Narrating the Destruction of the Body

Narrating violent acts raises unique questions about the ethics of different narrative forms. That does not mean that either phenomenology or narrative analysis imposes a code of ethics as something external to phenomena and stories and the relationality they describe. Rather, ethics is already inherent in the recognition of our intersubjective connectedness. Bearing that in mind, is it possible to make ethical judgments about the rhetoric and shape a story takes as well as the acts being portrayed? Since the forms of narratives—including the perspective from which a story is told, the time spent on different details, and the language used—invite specific ways of interpreting the violence being represented, I would suggest that it is possible. But the invitation issued by a text is only one element of a text's latent ethical potential. Readers are free to accept or reject the invitations offered by a text. We frequently have enough information about an event being portrayed to imagine alternative modes of portrayal. The reader's reception of a narrative is a complex and interesting field of inquiry, but for this chapter, I will set it aside to focus on the forms of the violent stories to be examined.[19] In the remainder of the chapter, I will look briefly at the three texts mentioned previously—one designed to encourage violence, one designed to encourage non-violence, and one that aims only to present for contemplation the manner in which violence exceeds, even annihilates, our sense-making faculties.

I will begin with the portrayal of violence in terrorist recruiting literature. Al-Qaeda began producing *Inspire*, a glossy propaganda magazine in English, in 2010 and ISIS followed with *Dabiq* in 2014. After the first 15 issues, ISIS was driven out of the Syrian town after which the magazine was named, so the ten subsequent issues have been published under a different title. The magazine is now called *Rumiyah*, or Rome. Disseminated on the web in English, the magazine includes articles about the interpretation and enforcement of Sharia law, narratives of military or terroristic activity and discussions of emigration and governance.[20] Although counter-terrorism scholar Thomas Hegghammer dismisses the magazine as insignificant in terms of the new information it reveals about ISIS's activities, the magazine's narrative framing is highly significant because it reveals narrative forms ISIS uses to elide the human suffering it wishes to cause.[21] With up to 77% of lone wolf terrorists being recruited through the internet, it seems wise to pay attention to the stories they read.[22]

On Tuesday, 22 March 2016, Najim Laachraoui and the brothers Khalid and Ibrahim El Bakraoui carried out a coordinated suicide bombing on the Brussels Zaventem Airport and the Maelbeek Metro station, near the EU headquarters. Thirty-five people, including the attackers, were killed, and 340 were injured.[23] Laachraoui and Mohamed Abrini, an accomplice later detained by police, had also been involved in the coordinated Paris attacks of 13 November 2015, which killed 130 people. After the attack on the Brussels airport, issue 14 of the ISIS magazine *Dabiq* featured a short article about it entitled "The Knights of Shahādah in Belgium," which provides biographies of Laachraoui and the El Bakraoui brothers. The "Forward" to the issue also celebrates the attacks and features photographs of the destruction at both the airport and metro station.

The bodies of terrorist "martyrs" and ISIS victims are frequently instrumentalized in the textual and visual media produced by Al-Hayat, the group's public relations company, and this issue of *Dabiq* is no exception, but the magazine begins in a different key. The "Forward," at first, portrays the body as vulnerable and as a site of the self's becoming within a community. The anonymous author writes:

> For nearly two years, Muslims in the lands of the Khilāfah have watched their beloved brothers, sisters, and children being relentlessly bombed by crusader warplanes. The scenes of carnage, of blood and limbs scattered in the streets, have become commonplace for the believers.[24]

Perhaps the most obvious point to be made here regarding the body is that the materiality of the body is foregrounded and presented in the expectation that readers will feel pity, disgust and anger at the reduction of "beloved brothers, sisters, and children" to "blood and limbs." The author uses the language of relationality to convey the sense that the entire community is wounded by the bombings and highlights the gap, impossible to bridge, between a beloved family member and his or her bodily remains. Because the body here is recognized in its fragility and interdependence, the forward at this point frames the body in a way that discourages violence.

Two paragraphs later, however, it becomes clear that only Muslim lives are framed as grievable, as Judith Butler would put it.[25] The author writes: "The death of a single Muslim, no matter his role in society, is more grave to the believer than the massacre of every kāfir on earth." Here, the categories of brother, sister and child cease to matter. One is either a Muslim or a "kāfir," a disbeliever. The body that served as the site of relationality and was incommensurable with mere remains is replaced by the symbolic political body: "Brussels, the heart of Europe, has been struck. The blood of its vitality spilled on the ground, trampled under the feet of the mujāhidīn." Transforming the body into a symbol eliminates the associations between embodied selfhood and life in a family and makes it impossible for the body to serve as a basis for understanding shared human vulnerability. As long as portrayals of hearts or limbs connote parts of the body that can feel pain, that can enable or impair our bodies' function, those portrayals evoke a shared experience and thereby form a basis for empathy. Once actual bodies disappear into symbolic bodies of Islam or Europe, that basis for empathy disappears. The mujāhidīn, or jihadis, keep their literal bodies ("feet of the mujāhidīn"), but non-Muslim bodies disappear into a political symbol.

The images paired with the forward include two photos of the wounded, one lying isolated on the ground, and the other receiving treatment for shock. In contrast, within the article itself, the magazine includes photographs of the ISIS operatives as healthy and in control. Ibrahim and Khalid El Bakraoui are pictured in headshots wearing t-shirts. Ibrahim is smiling. Laachraoui appears in an action shot. He smiles and winks at the camera while holding an AK-47. Behind him sits Mohamed Belkaid, who was suspected of involvement with the Paris attacks and then killed by police in a raid in Belgium. In the photo, Belkaid holds a bloody knife. In one way, the images accompanying the forward reverse the representation of the bodies referred to in that text. They show a "scene of carnage" like the one referenced by the text, and readers who do not comply with the text's separation of Muslims and non-Muslims into grievable and non-grievable victims can easily be led by the juxtaposition to think about the "beloved brothers, sisters and children" killed that day in Brussels. The images recall the real bodies that the symbolization of "the heart of Europe" had hidden. But if we compare the forward's images of the wounded in Brussels with the pictures of the ISIS operatives, then the images reinforce the text's implications. The categories of "mujāhidīn" and "kāfir" are maintained, obscuring attackers' and victims' shared humanity and shared embodied vulnerability.

Judith Butler points out in her *Frames of War* that "The 'frames' that work to differentiate the lives we can apprehend from those we cannot . . . generate specific ontologies of the subject."[26] The *Dabiq*'s forward generates Muslim subjects that are vulnerable, grievable and recognizable as objects of empathy because of their bodily fragility. If we read this back into an awareness of the always intersubjective nature of this subjectivity, then the lack of ability to perceive the embodied and relational life of a potential victim is most fundamentally the lack of the ability to see that person as already part of who we are, our love (Marion's term) or responsibility (Levinas's). Because *Dabiq* was published by a terrorist organization, it is not surprising to find here a narrative that enables violence by obscuring the universality of our shared, human embodiment, but the pattern of selectively individualizing victims and obscuring the intrinsic responsibility we have toward others is not hard to find in English or American mainstream media. It is a pattern that should disturb us not because of where I found this particular example, but because of the way it frames others as symbolic elements of an enemy force. The contrast between the instrumentalizing, heavily categorical portrayal of bodies here and the portrayal of bodies in King's "Letter" is striking.

Conceived as an open letter for *The Birmingham News*, King's "Letter from Birmingham Jail" addresses two main audiences—eight clergymen who had written a letter entitled "A Call for Unity" in the same newspaper, and readers of the newspaper who might be recruited to join protests and other acts of non-violent resistance. In association with the Alabama Christian Movement for Human Rights and the Southern Christian Leadership Conference, King had led the people of Birmingham in a series of boycotts, sit-ins and marches in the spring of 1963. Local police responded by arresting thousands of protesters. By the end of the first day of marches, 959 children between the ages of 6 and 18 had been arrested for walking from the 16th Street Baptist Church to the city hall. A few days later, another thousand were arrested.[27] The Commissioner for Public Safety, Eugene "Bull" Connor authorized the use of police dogs and high-pressure water hoses to bring marchers into submission after some protesters had thrown stones. The "Letter" was part of a coordinated publicity campaign to expose these abuses,[28] and it is now the "most widely reprinted" and "most widely read" document of the American Civil Rights era.[29]

King's "Letter" foregrounds a concern for bodily suffering comparable to the first quote discussed from *Dabiq*, only here the author imagines his own suffering at the hands of others. "We repeatedly asked ourselves," he remembers, "Are you able to accept blows without retaliating?"[30] Although it is his own bodily pain that he is imagining, he does not even obliquely let that suffering become a symbol of some greater social crisis or sacrificial offering. It is merely pain to be endured. Moreover, he does not privilege his own suffering over that of others, but defers the question to the anonymous "we" committed to non-violent protest in Birmingham. The symbolism of the body politic found in *Dabiq* recurs here. Segregation is a "disease" society suffers. But, the transference of bodily imagery into a symbolic field is never used to obscure the fragile physicality of real bodies.

For King, non-violent direct action requires a "crisis," and representations of the body undergoing violence lend essential energy to that crisis.[31] Picturing the crisis of civil rights as a crisis of the treatment of the body became, more generally, an effective strategy for the movement's presentation in writing, photography and film. For example, the photographs taken during Connor's crackdown on peaceful men, women and children galvanized the Civil Rights movement's attempts to gain widespread publicity. Capturing the moment of a dog's teeth near a man's belly through his tucked-in white shirt or a woman looking away as a high-velocity hose hits the man behind her, these images awoke international outrage.[32] King knew the power of these images, but his portrayal of the body in his writing is less sensational. He pictures the body as a site of shared human vulnerability in everyday settings as well as moments of crisis.

Late in the letter, King provides an extended description of bodily suffering, which is worth quoting at length:

> Perhaps it is easy for those who have never felt the stinging darts of segregation to say, "Wait." But when you have seen vicious mobs lynch your mothers and fathers at will and drown your sisters and brothers at whim; when you have seen hate filled policemen curse, kick and even kill your black brothers and sisters; when you see the vast majority of your twenty million Negro brothers smothering in an airtight cage of poverty in the midst of an affluent society; when you suddenly find your tongue twisted and your speech stammering as you seek to explain to your six year old daughter why she can't go to the public amusement park that has just been advertised on television, and see tears welling up in her eyes when she is told that Funtown is closed to colored children, and see ominous clouds of inferiority beginning to form in her little mental sky, and see her beginning to distort her personality by developing an unconscious bitterness toward white people . . . then you will understand why we find it difficult to wait.[33]

King's use of "you" here invites the reader to imagine this suffering occurring to someone he/she loves, a reminder of the fundamental connectedness of human beings. He describes the body, not as an object, but as the mediator, given without choice in black and white, between the world hostile to his daughter and her growing consciousness. The specificity of the experience he records as taking place with "your daughter" is clearly not the experience of a white reader's daughter, but King contests the categorical construction of race by writing in confidence that one person can substitute the vision of tears in this black daughter's eyes with any daughter's suffering. The possibility of that absorption into another's suffering is made possible by the shared bodily sign of tears. Throughout the letter, King reminds readers that what people on both sides of the then broiling segregation debate live through and suffer through is what Marion calls the "flesh," not the unfeeling body, and he emphasizes that what we experience via our bodies is all consciousness has to work with and is often given by factors outside of our control. He writes within incarnation as extreme passivity.

The rhetoric of bodily suffering overlaps somewhat with the rhetoric of *Dabiq*, but to much different effect. In King's text, the evocation of family ties carries the same or greater emotional force than the mention of beloved brothers, sisters and children in *Dabiq*. Although the suffering described in *Dabiq* is more physically extreme, King's daughter's suffering is described within a specific setting. She is individualized through the details of age and specific six-year-old desires. The text resists making her representative of an age, race or region, through these individualizing details, but emphasizes that this particular little girl, like every other particular little girl, is the reader's responsibility. King's text also resembles the forward in *Dabiq* by emphasizing that the body is not mere materiality, but the center of a self's lifeworld with its own "mental sky." King asserts, however, that no one, regardless of the suffering they have to avenge or the categories with which society limits them has the right to harm another's body. He argues, in fact, that they cannot harm someone else without damaging the network of intersubjective relations of which they are already a part.[34]

Finally, Joseph Conrad's *The Secret Agent: A Simple Tale* (1907) centers on a failed terrorist attack in which an explosive device accidentally detonates, killing a cognitively impaired young man who had been asked to carry it by his guardian, his sister's husband. His main character, Adolf Verloc, works as an agent provocateur in London and, as a cover, runs a shop selling "shady wares" and "photographs of more or less undressed dancing girls."[35] He lives in the house above the shop with his wife, Winnie, her mother, and her cognitively disabled younger brother, Stevie. In the evenings, the family is joined by Verloc's anarchist associates. In the novel's second chapter, readers learn that Mr. Vladimir, the First Secretary of the embassy employing Verloc, is not pleased with his work and that he must carry out a special mission—bombing the Greenwich Observatory and framing the anarchist cell—in order to maintain his position. In order to accomplish this mission, Verloc obtains explosives from The Professor, one of the anarchist characters, and conceals them in a brown-paper package. He then invites Stevie, who dotes on him, to accompany him for a walk. He has shown Stevie the direction to walk into Greenwich park and out again. The fourteen-year-old boy is supposed to put the brown package down inside the park and then rejoin his brother-in-law, but he trips and the package explodes prematurely.

The remains of Stevie's body are described in crassly material terms after the explosion. They are "mingled with things that seem to have been collected in shambles and rag shops," and collected with a shovel.[36] The body appears in the text as evidence at first, more than a human being to be mourned, so the effect of describing the body as merely material is quite different than in *Dabiq*, where the damaged limbs are linked in the same sentence to beloved family members. In *The Secret Agent*, three and a half pages are devoted to the investigation of "a heap of nameless fragments" between two waterproof sheets, with no one in the room aware of who the body belonged to.[37] Although readers, by this point in the novel, already associate Stevie with "blind love and docility," and although we have seen his mother and sister prioritizing his needs over their own, we cannot feel grief over his reduction to a heap of fragments because the body is unidentified.[38] The reduction of the person to a mere material body becomes nauseating rather than tragic. Other bodies are also spoken of throughout the novel as though they were not feeling flesh, but objects. Verloc's body is referred to as a "mortal envelope" when he is dead and alive, suggesting an instrumentalization of the body for the sake of whatever immortal message he might have had to deliver.[39] Lips, ears and noses are described in grotesque detail, not as organs for mediating experience in the world, but according to shape and texture.

The material presentations of characters' bodies are made forceful through repetition. This is particularly true of Stevie's body. And yet, the novel confronts readers with the inadequacy of this presentation even as it refuses to describe bodies in any other way. Conrad never portrays what Stevie knew or saw or felt at the moment of his explosion, but he narrates both Chief Inspector Heat, who is investigating the explosion, and Winnie Verloc, Stevie's sister, trying to imagine what Stevie had experienced.[40] Inspector Heat is the more imaginative of the two. "No physiologist, and still less of a metaphysician," he

> rose by the force of sympathy, which is a form of fear, above the vulgar conception of time. Instantaneous! He remembered all he had ever read in popular publications of long and terrifying dreams dreamed in the instant of waking; of the whole past life lived with frightful intensity by a drowning man as his doomed head bobs up.[41]

He does not know enough about Stevie to imagine more, and the reader does not yet associate these remains with the debilitatingly compassionate young man we get to know throughout the book, so we cannot fill in the inspector's gap in knowledge. In short, there is no illusion that either readers or Inspector Heat learn anything about Stevie's last moments; it is the act of trying to imagine those moments that stands out. In a book where characters assume knowledge about one another based on popular "scientific" typologies, where people are repeatedly described according to bodily abnormalities rather than personal history or character trait, Inspector Heat's "force of sympathy" shines.

The only other character who consistently demonstrates an ability to sympathize with physical suffering is Stevie himself. In a conversation about burning the skin of prisoners, the eyes of the anarchists conversing are described as "extinguished," "vacant," black and "blind."[42] But Stevie's eyes are "scared." They "blaze with indignation." The narrative slips into free-indirect discourse and adopts the boy's simple vocabulary: "hot iron applied to one's skin hurt very much . . . it would hurt terribly."[43] Although his vocabulary has none of the refinement of the men he listens to, Stevie knows more. Through him, Conrad portrays our ability, via prior bodily experience, to imagine another's suffering and react on behalf of the body of the other. Stevie experiences the woundedness of a horse struck by a whip, a cabman with a prosthetic hook and the undifferentiated prisoner of this conversation, showing how capacious this embodied empathy can be, but Conrad suggests how common it is for the natural awareness of these Levinasian intersubjective wounds to be deadened by perception of material bodies rather than feeling flesh.

Through Inspector Heat's reflection on the body before him, Stevie's compassion and the novel's overall narrative structure, *The Secret Agent* presents violence as a senseless and unpredictable product of cruelty. Generalized cruelty. Responsibility for Stevie's death must be shared by Mr. Vladimir, who ordered the bombing; by Verloc, who put the bomb in the boy's hands; by Winnie Verloc, who suggested Mr. Verloc take Stevie on more walks and by The Professor, who made the bomb. At the moment Winnie imagines the bomb exploding in Stevie's hands, she plans to kill her husband, a plan she carries out minutes later with the carving knife they had used at dinner.[44] Stevie, however little he might understand of his mission, is a terrorist. Nobody in the book is innocent. Conrad asserts in his "Author's Note" that the novel's most central story is the story of Winnie Verloc, whose sympathy for one person blinds her to suffering and fragility of another.[45] That is a pitfall for the reader as well, since Stevie and Verloc, the characters who become victims, are also the most developed. But the strong contrast between the book's overwhelming presentation of the body as material and the few instances of body-based empathy points to a more general tragedy that can be expressed in phenomenological terms. We are involved via the feeling flesh in the suffering of others and what appears to be an individual act of violence (the murder of Stevie or Verloc) has its real source in a general, cultural obfuscation of that fact.

Conclusion

The most central question I have tried to answer in this chapter is how narrative representations of the body and especially other people's bodies deaden what Levinas describes as a fundamental pre-conscious recognition of our mutual dependence, our mutual responsibility and vulnerability. The one-sided elaboration of relationality and consciousness, as we saw in *Dabiq* (and as we see often enough in Western European or American media) seems to be used to this effect. The transference of suffering and consciousness from an individual flesh onto a symbolic whole is similarly problematic. The material presentation or typological categorization of the body featured in *The Secret Agent* can facilitate the contemplation of shared embodiment and sympathy, as with Inspector Heat and Stevie, but if the body's materiality becomes the primary focus rather than a moment of exceptional realization, as was the case with the anarchists and much of the novel's third-person narration, then it has the opposite effect. Bodies are seen instead of persons.

We can also approach this question the opposite way—how can narratives ethically portray the violation of personhood and shared life that violence enacts? Reflection on the fragility of the body appears to be necessary but not enough. A text like King's "Letter from Birmingham Jail" couples that fragility with a sense of responsibility for the other. King's text focuses on the positive potential of that responsibility, its power to effect social change, while Conrad's text focused on the negative potential of that responsibility, the way its neglect makes people complicit in violence. In both cases, however, the ultimate effect is to reinforce the way one individual's self-becoming inevitably involves the suffering of others. The other feature that both King's "Letter" and Conrad's novel share is the distinction between the material body and the feeling flesh. The *Körper* and *Leib* distinction is familiar to phenomenologists, but within a narrative framework the distinction's ethical potential is revealed as some people capable of violence are portrayed as ignoring the feeling flesh.

As Levinas writes, "Violence does not stop discourse."[46] It is a part of it. The person performing violence interprets another person as worthy of destruction on the basis of discourse. Through the violent act itself, some message is conveyed although no message can ever exhaust what is lost in the destruction of another, vulnerable person. By working within and through already-operating discourses, narratives portraying political violence have the potential to contest understandings of human relationality that make violence seem permissible or even good. People are moved to action by stories. At times, stories are explicitly persuasive, as with "Letter from Birmingham Jail" or *Dabiq*. In other cases, as with *The Secret Agent*, stories change readers by giving us new ways of understanding the body's place in our relationships to the world and other people. Like our bodies themselves, stories are something that we are always already within, something that we see ourselves and others through without always seeing the story (or the body) as a means of mediation. The relational phenomenology of Levinas and Marion helps clarify the ethical implications of stories about political violence by recalling the foundational interconnectedness that stories reinforce or conceal.

Notes

1. EEAS Strategic Planning, *A Secure Europe in a Better World: European Security Strategy Report* (December 12, 2003). (Brussels: European Union), 2.

   Marcel Graça, "Promotion and Protection of The Rights Of Children Impact Of Armed Conflict On Children," (August 26, 1996). Report to the UN General Assembly. 9.

   This statistic has been debated by some scholars. The most comprehensive critique is this: Adam Roberts, "Lives and Statistics: Are 90% of War Victims Civilians?" *Survival* 52, no. 3 (2010): 115–36. Roberts questions whether deaths from "indirect effects of war such as disease, malnutrition and lawlessness" should be included in statistics of civilian war deaths and asks that statisticians distinguish between those "who die as a direct effect of war" and those who "even after the war is over, die prematurely from injuries sustained in war" (116). While this is a valid statistical concern, I would argue that these other deaths should be considered in measurements of civilian war death. For an account of the difficulties of these calculations, see: Jeff Lewis and Belinda Lewis, "The Myth of Declining Violence: Liberal Evolutionism and Violent Complexity," *International Journal of Cultural Studies* (January 2017): 1–17.
2. Steven Pinker, *The Better Angels of Our Nature: Why Violence Has Declined* (New York: Viking, 2011).
3. For a history of the ideological debates about killing civilians and an overview of current practices, see: Hugo Slim, *Killing Civilians: Method, Madness and Morality in War.* (Oxford: Oxford University Press, 2010).
4. Michael Staudigl, "Racism: On the Phenomenology of Embodied Desocialization," *Continental Philosophy Review* 45 (2012): 23–39, 28.
5. Joseph Conrad, *The Secret Agent* (Oxford: Oxford University Press, 2008), 123, 212.
6. Judith Butler, *Precarious Life: The Powers of Mourning and Violence* (New York: Verso, 2001), xviii.
7. Judith Butler, *Frames of War: When is Life Greivable?* (New York: Verso, 2009), 77.
8. Michel Henry, "Phenomenology of Life," *Angelaki Journal of the Theoretical Humanities* 8, no. 2 (June 2010): 97–110, especially 103. DOI: 10.1080/0969725032000162602; Butler, *Frames of War*, 95.
9. Zygmunt Bauman, *Collateral Damage: Social Inequalities in a Global Age* (Cambridge: Polity, 2011), 59.
10. Zygmunt Bauman, *Postmodern Ethics* (Oxford: Blackwell, 1993), 53.
11. Michael Staudigl, *Phenomenologies of Violence* (Leiden: Brill, 2014), 9.
12. Maurice Merleau-Ponty, *The Phenomenology of Perception*, trans. Colin Smith (London: Routledge, 2002), 94–5.
13. Jean-Luc Marion, *In Excess: Studies in Saturated Phenomena*, trans. Robyn Horner and Vincent Berraud (New York: Fordham University Press, 2002), 86–7.
14. Levinas recounts this idea many places. One very clear and succinct instance is Emmanuel Levinas, *Alterity and Transcendence*, trans. Michael B. Smith (London: The Athelone Press, 1999), 104. Marion, *In Excess*, 115–16.
15. Emmanuel Levinas, *Otherwise than Being, or Beyond Essence*, trans. Alphonso Lingis (Dordrecht: Kluwer Academic Publishers, 1991), 195, note 12.
16. Ibid., 95.
17. Jean-Luc Marion, *The Erotic Phenomenon*, trans. Stephen Lewis (Chicago: University of Chicago Press, 2007), 120–1.

18. Jacques Derrida, *The Work of Mourning* (Chicago: University of Chicago Press, 2001), 198.
19. For a phenomenological ethics of reception, see Cassandra Falke, *The Phenomenology of Love and Reading* (New York: Bloomsbury Press, 2017).
20. Table 1, in Brandon Colas's "What Does Dabiq Do? ISIS Hermeneutics and Organizational Fractures within Dabiq Magazine" features a more detailed breakdown of the magazine's content. *Studies in Conflict & Terrorism* 40, no. 3 (June 2016), 173–90, doi.org/10.1080/1057610X.2016.1184062
21. Thomas Hegghammer, *Un-Inspired* (Jihadica: Documenting the Global Jihad) (July 6, 2010).
22. Kareem El Damanhoury and Carol Winkler, "Picturing Law and Order: A Visual Framing Analysis of ISIS's *Dabiq* Magazine," *Arab Media and Society* (February 15, 2018) www.arabmediasociety.com/picturing-law-and-order-a-visual-framing-analysis-of-isiss-dabiq-magazine/
23. BBC News, "Brussels Explosions: What We Know About Airport and Metro Attacks" (April 9, 2016).
24. ISIS, "Forward," *Dabiq* 14 (1437 Rajab; April 2016), 4–5. All issues of *Dabiq* and *Rumiyah* are available through The Clarion Project website: clarionproject.org
25. Butler, *Frames of War.*
26. Ibid., 3.
27. Steven E. Barkan, "Legal Control of the Southern Civil Rights Movement," *Law and Social Movements*, ed. Michael McCann (New York: Routledge, 2006), 412–3.
28. S. Jonathan Bass, *Blessed Are the Peacemakers: Martin Luther King, Jr., Eight White Religious Leaders, and the "Letter from Birmingham Jail"* (Baton Rouge: LSU Press, 2001).
29. Maralyn DeLaure and Bernard K. Duffy, "Martin Luther King, Jr. 1929–1968," in *American Voices: An Encyclopedia of Contemporary Orators*, ed. Bernard K. Duffy and Richard W. Leeman (Westport, CT: Greenwood Press, 2005), 258–69, 262.
30. Martin Luther King, "Letter from Birmingham Jail," in *Why We Can't Wait* (New York: Harper & Row, 1964), 80.
31. Ibid., 81.
32. Diane McWhorter discusses photographs from the Birmingham campaign at length in *Carry Me Home: Birmingham, Alabama, the Climactic Battle of the Civil Rights Revolution* (New York: Simon & Schuster, 2013). The two images referred to here are Bill Hudson's photograph of 17-year-old Walter Gadsen being jumped on by police dog Leo, and Charles Moore's "Birmingham Protests, 1963," which was published by *Life* magazine. Effective though it was, Hudson's photograph is misleading. Gadsen, who was not protesting, actually caught the dog with his knee and ended Leo's career.
33. King, "Letter," 83.
34. See, for example, Martin Luther King, *Where Do We Go from Here: Chaos or Community?* (Boston: Beacon Press, 1968), 64.
35. Conrad, *The Secret Agent*, 3–4.
36. Ibid., 65, 154.
37. Ibid., 65.
38. Ibid., 8.
39. Ibid., 193, 28.
40. Ibid., 65, 191.
41. Ibid., 65.
42. Ibid., 35–6.

43. Ibid., 36.
44. Ibid., 193.
45. Joseph Conrad, "Author's Note," in *The Secret Agent*, ed. John Lyon (Oxford: Oxford University Press, 2008), 231.
46. Emmanuel Levinas, *Totality and Infinity: An Essay on Exteriority*, trans. A. Lingis (Pittsburgh, PA: Duquesne University Press, 1969), 239.

Section II
Suffering Bodies

5   Only Vulnerable Creatures Suffer

On Suffering, Embodiment and Existential Health

Ola Sigurdson

In this chapter, I wish to explore how suffering, embodiment and existential health relate to each other through a phenomenological analysis. More precisely, as I am convinced that such phenomena are both experienced and communicated through words (although not exclusively), my inquiry could be described as phenomenological hermeneutics. I am especially interested in how suffering, which I will argue should be distinguished from the experience of pain, relates to human agency, and how suffering then becomes a particular mode of being in the world. Suffering, I will argue, is essential for the understanding of human beings as vulnerable creatures. It is through suffering as an active passivity that we can transform the experience of pain (or any cause of suffering) to a constructive relationship that could be called *existential health*. But this understanding of suffering as a particular mode of being in the world and the corresponding notion of existential health are not immediately given in and through some uninterpreted experience, but as historical and linguistic phenomena. Thus my emphasis on a *hermeneutical* phenomenology, and a critical hermeneutics at that, that wishes to acknowledge that any experience of suffering is dependent upon language and therefore on culture; likewise is any meaning of suffering only given through interpretation.[1]

My exposition of the relationship between suffering, embodiment and existential health will proceed in three parts, on suffering, embodiment and existential health, followed by a short conclusion.

Suffering

What does the term *suffer* mean? According to the *Oxford English Dictionary* there are a number of different meanings attached to the word, but not all of them are immediately significant from a phenomenological perspective, so let me stay with just a few of them. To begin with, *suffer* can mean, in the intransitive use of the word, "to undergo pain, punishment, or death," or, in its transitive use, to experience "a disease or ailment." The first use is exemplified by the short statement "I suffer." If you tell me that you are suffering I would, in most cases perhaps, draw the conclusion that you are in pain, either mentally or physically or both. Likewise, in the sentence "a brave man suffers in silence." In its transitive use, we suffer a particular disease or ailment, as if I would state that "I am still suffering from a headache." In everyday parlance, the word *suffer* probably often equals *pain* or *disease*. But, interestingly enough, there is another meaning, which the *OED* characterizes as "now rare," namely "to be the object of an action, be acted upon, be passive." Suffering, in this sense, is not necessarily to be equal with pain or ailments, but is rather the opposite of agency, which is more obvious in the complementary antonyms passive and active.

The term *passivity* is related to *passion* which is a word that in many ways functions in a similar way to *suffering*, even if it is either more erotically or more theologically charged, or both. To undergo a passion is to be swept away by emotions evoked by a significant other (or a significant phenomenon or thing). To endure suffering is, or could be, to imitate Christ's sufferings on the cross. Indeed, the term *passion* in English is borrowed from Latin, *passio* in nominative singular, where the primary connotations come from Christian theology. In the passion narratives of the four Gospels, Christ, the Son of God, is portrayed as enduring, suffering, undergoing the flogging, public humiliation, crucifixion and death in a way that has been paradigmatic for the historic social imagination not only of what it means to suffer but also for the passivity that is involved in human existence. Of course, in the passion of Christ, *passion* obviously involves enduring extreme pain. But it also means, and in a way that I would suggest is distinct from the feeling of pain, to be passive.

Passion, today, also means *strong affection* or *sexual desire*, where the passive dimension of the mental state is at least implicit in the sense of being overwhelmed by the affection or desire in question. Maybe it is even more correct to speak of being overwhelmed by the object of my affection or desire in such a way that I experience a certain forgetfulness of myself as subject. To speak of "suffering a passion," then, would not be pleonastic, if *passion* directs our attention to the object-pole of the mind's intentionality and *suffer* the subject-pole. To "suffer a passion," then, is not only to be directed toward the object of my attention or desire, but also to more or less consciously relate to this standing out of myself (*ek-stasis*) as something I am enduring or undergoing. This is of course a matter of nuance, but this way of understanding what it means to "suffer a passion" seems to introduce a more complex notion of subjectivity, where the suffering of my passion, at least when formulated through verbs like *enduring* or *undergoing*, announces some minimal but active agency in relation to being swept away by the object of my desire. If I *suffer* my passion rather than just *being passionate*, then there is a distinction between being passionate and suffering that refers to a relation of the subject to this passivity. If there is a relation, then the passion is not only a blind intervention from outside, from an object that dazzles us unconditionally and freezes our perceptive faculties. To put it in other terms, there is then a question of subjectivity, since I can call this passion mine, as something that belongs to me intrinsically.

The point of my teasing out different nuances of *suffering* or *passion* is not to create a philosophy from etymology or the use of words, but rather to describe the experience of suffering in a way that might be of phenomenological interest. The erotic dimension of passion will not play any important role in my discussion in this chapter, except as a background to the more extensive analysis of suffering as undergoing pain or other less pleasant experiences. What I have suggested so far is that suffering as an experience, irrespective of if it involves pain or ailment, carries with it an aspect of passivity. This passivity, however, is not just an extrinsic receiving, like a glass that receives water when we pour ourselves a drink. It is a passivity that involves us as subjects; we are the subjects of the experience of suffering; it does not cancel our agency. Consider, once again, the experience of pain. It is perfectly plausible to speak of *suffering pain* in the sense of enduring pain, and such a suffering seems to imply a certain reflexive relation to the pain. It is not just that I am in pain, but somehow I relate to this pain in suffering so that there is a distance between me and the pain. Even if an extreme pain is thinkable that would force me to express not only that "I am *in* pain" but also "I am pain," thus suggesting an identification between me and the pain, this still would imply a minimal distinction by necessity as the copula *am* suggests. Suffering as an experience would consequently be the seemingly paradoxical experience of an active passivity; *seemingly*, I think, since there is an intrinsic difference between the passivity of a human being and the glass into which we pour our drink. Even if suffering is an experience of passivity, it does not necessarily cancel the agency of the human being in question. Pain could very well cancel our agency, I suppose, by being so intense that we lose consciousness or die, but suffering belongs to the agency of the person. A human being is an agent that has the capability to suffer.

Is this necessarily so? Even if there is reason to speak of agency even in the experience of suffering, is not the volition involved so minimal that it makes no sense to speak of agency? In the experience of suffering, the involuntariness is fundamental or else it would not be suffering in the sense of undergoing or enduring. If I could choose to feel the pain or not, if I could choose to be overwhelmed by the object of my affection or not, if whatever is the cause of my suffering would not be able to catch me unaware, would there be a sense of talking about suffering at all? There is, in our human existence, sometimes a tendency to try to become invulnerable, so as not to be overwhelmed by pain, love, desire or chance. This is by no means a bad thing, necessarily, but in the extreme it would be an abdication from a common understanding of human beings as vulnerable creatures. In some sense vulnerability seems to be an essential human trait, especially in our existence as embodied beings, and is the prerequisite for experiencing suffering. Only vulnerable creatures suffer. Invulnerable creatures are in control, but vulnerable creatures are exposed to the contingencies of existence in space and time. Something could actually happen to me in a way that affects me intrinsically.

If, then, all human beings are vulnerable and therefore prone to suffering in some sense, the question still is whether this vulnerability is a characteristic trait of the humanity of human beings or not. According to historian Judith Perkins, suffering sometimes fell outside of the representation of the humanity of human beings in late antiquity; not that they did not suffer pain, but this was not understood as essential to their humanity. The true representation of what it means to be a human being is to keep one's agency, whereas the passive reception of pain fell outside of the domain of the truly human. To paraphrase one classic philosopher: where I am, suffering is not, and where suffering is, I am not (Epicurus actually spoke about death, not suffering). Suffering and vulnerability were not seen as essential to humanity; no doubt pain was experienced even by philosophers from late antiquity in no small measure, but pain would be conquered by suicide rather than endured through suffering as an affirmation of agency as the essence of humanity. Epictetus, as one of Perkin's prime examples of this attitude, emphasized self-control and mastery and denied the importance of anything that would affect the subject from outside. As Perkins puts it, for Epictetus it was all about control:

> He instructed his students in techniques for mastering and controlling their attitudes and feelings to ensure that their equanimity could never be disturbed. He taught them to construct a self immune from desire, pain, grief, or fright. The basis of his teaching was a detachment from everything outside the control of the mind. This included the body.[2]

It is through the Christian martyrs such as Ignatius of Antioch as well as the Greek physician Galen that both the possibility and the actuality of suffering came to be a part of the understanding of genuine subjectivity. This did not mean that they made a virtue out of a necessity, as Christianity was a persecuted minority at the time. As Perkins writes, "[t]he martyr *Acts* refuse to read the martyr's broken bodies as defeat, but reverse the reading, insisting on interpreting them as symbols of victory over society's power."[3] Rather, it entailed a redefinition of subjectivity through changing the representation of the human self. Through this redefinition of subjectivity or even reversal of the understanding of it, suffering is no longer seen as something external to the self but rather internal to it. Self-definition is achieved, not through *avoiding* suffering but *through* suffering. The martyr is not only someone who suffers pain, but a witness (*martureo* in Greek means "to bear witness" or "testify") and someone who testifies precisely through his or her suffering.

The effects of this reversal came to have a profound cultural impact on the representation of pain and disease as well as the role of embodiment for human self-understanding. No doubt this historic account would need to be elaborated in more detail; what I want to suggest through this historic excursus, however, is the more theoretical point that suffering and vulnerability as essential traits of human self-definition are not given in the sense that they are part of every possible philosophical or theological attempt to characterize what it means to be a human being. There is indeed an ongoing struggle whether to understand the human self as an essentially vulnerable creature or an agent that is in control no matter what. Formulated in another way, how agency and passivity are related to each other is an issue, not only historically, where Christianity came to emphasize the significance of passivity over against the classic accent on active agency. It is also a question that returns in philosophical anthropology, gender studies, transhumanism, even in humor theory, but that is another matter at the moment. A phenomenological account of vulnerability and suffering such as mine will in all likelihood not serve as a final argument on behalf of passivity. Even when a phenomenological analysis would show how suffering unavoidably belongs to an embodied sentient and thinking creature by virtue of its very embodiment, this would not amount to any *proof* that vulnerability is a necessary dimension of what it means to be a human being, only what would be involved in affirming or denying that such is the case. The meaning or location of suffering in human existence is a matter of mode of life and world-view in conjunction.[4]

Embodying Suffering

What does this mean for the bodily conditions of suffering? The body, from a phenomenological perspective, is not a particular thing or an object, as in some (modern) philosophical traditions, but rather the way we as human beings live in the world and relate to others as well as ourselves. To use a phrase that has been repeated many times in phenomenology, we don't *have* a body so much as we *are* a body. This does not mean that the body is not *also* a physical object, but to understand what it means to be a physical body, it is not enough just to point to its physicality even if this physicality is an important aspect of our embodiedness.

According to the French philosopher Maurice Merleau-Ponty, who is one of the more well-known among many philosophers dealing with embodiment in the phenomenological tradition, we need to make a distinction between the objectified body and the more original phenomenal body. One way that Merleau-Ponty understands the phenomenal body is the following: "The body is the vehicle of being in the world, and having a body is, for a living creature, to be intervolved in a definite environment, to identify oneself with certain projects and be continually committed to then."[5] The phenomenal body is the horizon of the objectified body, and to understand what embodiment means before and beyond the representation of the body as an object, we need to engage with the phenomenal body. This phenomenal body is characterized by spatiality and mobility, and should be understood more as the possibility to orient ourselves in the world. Just to take one mundane example borrowed from Merleau-Ponty: to learn the art of typing is neither something intellectual nor something instinctive, but something that needs to be trained until it becomes a habit. When this would-be habit finally becomes a proper habit that will come spontaneously when called for, then it belongs to our repertoire of possibilities and as such has widened the horizon for our being in the world. The physicality of the body is of course instrumental in our ability to type in some very fundamental way, as we need to press the keys, but the ability is not reducible to our fingers stroking the keys. If I lost my fingers in an accident, my phenomenal body would still retain the ability to type even if I would not be able to actually execute it. Further, as Merleau-Ponty remarks, to be able to type does not mean that we are able to explain, in an abstract or intellectual way how to type. The ability to type is a knowledge inscribed as a possibility in the repertoire of our phenomenal bodies. Whether we possess it or not, it is only one of many embodied possibilities, ranging from the most mundane and almost instinctual, like walking or speaking, to the more profound and artistic, like dancing or playing football like Lionel Messi. The phenomenal body is performative, to use an attribute borrowed from Judith Butler, rather than a thing. Phenomenologically, typing is an embodied ability perhaps not even primarily related to our motoric skills as such, but to feeling, thinking, fantasizing and so on.

There is indeed a lot more to be said about Merleau-Ponty's conception of the phenomenal body as well as philosophies of the body in general, but let me now move forward to an account of embodiment that will begin to explain how one could understand the bodily conditions of suffering. Even when he is arguing against the mind-body dualism that is to be found extensively in the philosophical tradition after Descartes, the philosopher and doctor Drew Leder has suggested that there is an experiential basis for such a dualism in the experience of pain and sickness.[6] Arguing from an account of embodiment similar to the one I briefly presented earlier, Leder suggests that there are modes of bodily "absence" where the body withdraws from our experience. Of interest here is foremost the kind of absence that Leder calls *dys-appearance* after the Greek prefix *dys-* which means bad, harsh or sick. In the particular form of absence where the dys-appearing body withdraws from us, it is not a matter of a forgetting of the body but rather the opposite, the experience of the body as something alien to our purposes and plans. The body in pain or the sick body appears to us for its own sake, not as a vehicle for carrying out our actions.[7]

When I happily type away on the keyboard of my computer, I don't extend any thought to my fingers as I have already mastered the art of typing; even as a precondition for typing my fingers remain unthematized by my consciousness. I am, for instance, focused on the screen to see the words of my thinking. But in a certain sense, it is not the screen I look at nor the actual lines of text but my thinking as such. In a way, I am thinking through my fingers and my computer. My intentionality is directed toward my thinking, neither—in this example—my typing nor the computer. This is as it should be, since the relative invisibility or anonymity of my fingers belongs to the art, and means that I am no longer in the process of learning this skill. But imagine a pain in one of my fingers, and the typing becomes interrupted by the finger suddenly appearing to my consciousness in its own right. Maybe it hurts so badly that I am unable to write even though I want to. This means that my body is experienced as in one way alien to me; it stands in the way of what I currently want to execute through the abilities of my phenomenal body. I know how to type. I want to type because I need to finish an assignment on time, but I cannot, because my finger hurts. This is an example, albeit somewhat trivial, of what Leder means by the dys-appearance of the body. The dys-appearing body is the body that in one the same movement becomes present to me but present just as something alien to me. It is no longer experienced as something identical to my own self, but somehow it is like something other is working through it. It questions the uninhibited function of my body in typing.

There are other ways in which the body is absent or alien than in its dys-appearing, but this is the way of most importance for my purposes here. My example of the sudden pain in a finger while typing is for illustrative purposes only. As we all know, there are indeed worse experiences than that, experiences that somehow inhibit our purposes and where we experience our body as working against us. Being embodied means to suffer from physical fragility, perishability and finitude: there are indeed pains and diseases that will put a stop to our abilities and ambitions in more profound ways than this. The dualism between mind and body is indeed something congruent with some possible experiences, and to understand suffering in relation to embodiment we would need to accommodate this experience in our philosophy without succumbing to an ontological dualism. What the experience of pain and the dys-appearing body means from the perspective of a phenomenological understanding of the body is that there opens up a gap within our experience of our bodies. Even if it is indeed true that we *are* bodies rather than *have* bodies, maybe we still do *have* bodies in an authentic sense, namely as the experience, within our embodied lives, of the body as a burden. To suffer is indeed an embodied experience, but perhaps an experience of precisely this gap, where we undergo or endure pain as something that both belongs to us and doesn't.

One of the shortcomings of a Cartesian dualism when it comes to suffering is, I think, that it makes the relation between the body in pain and the suffering awareness of that pain extrinsic. It is not really me who suffers pain but my body. The perspective on embodiment I suggest here would hold on to the ambivalence of the experience of the body in pain: my subjectivity here is found in the experience of both having and being a body at the same time, not just on the side of the consciousness. In his twentieth seminar from 1972 to 1973, *Encore*, the French psychoanalyst Jacques Lacan claims that there is a language of the body that does not coincide with consciousness: "I speak without knowing it. I speak with my body and I do so unbeknownst to myself. Thus I always say more than I know."[8] The subject in analytical discourse, according to Lacan, is the *I* that comes to expression in what is said without itself being aware of itself. Subjectivity is not identical to the Cartesian *res cogitans*; intentional consciousness is not the master of its own house. What Lacan is referring to is a subjectivity whose conscious dimension or intentionality is entangled in embodiment in a way that will not allow any neat separation between mind and body. In the same way that there is no naked encounter between *I* and *Thou* beyond or outside of the symbolic context, there is no mind and body outside of our symbolically mediated embodiment. Subjectivity is not just anchored in *I think* but also in a body that continually exceeds itself through itself. The reflexive ability then, which in some sense is a necessary condition for subjectivity, is not just a matter of thinking or something that has an extrinsic relation to our embodiment, but lies inherent in our bodily constitution as such. Thus, my experience of my own pain allows for a distinction between my self and the pain, but this distinction does not revoke the fundamental solidarity between the two.

I deliberately chose the term *solidarity* to describe the relation between my self and the pain, as this value-laden word points to the intrinsic relation between the two. As I argued earlier, the experience of suffering is an experience of passivity, and a passivity that is in some sense inevitable, which means that the solidarity between me and the pain is shown by its spontaneous nature. I don't have to decide to acknowledge my pain as mine, but it is given as mine. That there might be ways of training oneself to increase the distance between the pain and myself does not contradict the basic givenness of this solidarity but rather shows it even more clearly. This solidarity is not only a basic tenet of one's own pain but also comes into play in relation to the pain of other beings. There is indeed no contradiction in saying that we suffer each other's pains. This does not mean that we feel the actual physical pain, but rather that the sense of another's pain spontaneously overwhelms us and becomes an accompanying trait of our existence for some time. I can feel in my body somehow that the other is in pain. There are of course a number of reservations and qualifications that would be needed here to make this last claim more substantial, but suffice it to say that we actually do experience a passive solidarity through suffering another's pain.

My own pain or the other's pain is addressed to me, I am the recipient of the experience of pain, and the solidarity between me and the pain means that I cannot but acknowledge the reception of the experience. The pain overwhelms me, regardless of whether I am prepared or if it catches me unaware. I receive it in my body, not exactly the pain which is already there, but the suffering through which I acknowledge the pain as mine despite its alienating effects of my embodied life. Through suffering in my body, I gain a relation to the pain, which although based on a passive impression allows me to transform this into an active passivity. Suffering thus entails a certain transformation or perhaps even transfiguration of the pain in that I gain, not a mastery over it perhaps, but at least such a relation that allows me both to encounter and confront it. This brings us to the question what health might mean in the light of embodied suffering.

Existential Health and Suffering

What kind health might be involved in suffering? There are many different definitions of health, and a good starting point is often the WHO definition of health from 1948: "Health is a state of complete physical, mental and social well-being and not merely the absence of disease or infirmity." One understanding of health, then, would be the absence of pain, at least involuntary pain, either physical or mental. This is something I believe is both legitimate and crucial to strive for, and when I am suggesting that suffering and vulnerability might belong to the conditions of human existence, I am not advocating some kind of heroic ethos that voluntarily tries to endure pain for its supposedly redemptive or character-building properties, quite the contrary. But when infirmities or challenges or losses are of a more persistent kind, there is no health care in this sense that can take care of these in the same way that a painkiller might take away my headache. Given that we live a life where we by necessity are not always in control, there are at all times possible events that might occur and overwhelm us, engaging us in a way that we have no way of avoiding. Such events might involve physical pain, but also loss and bereavement, loneliness, lack of meaning and trust, disease and death and so on. For many people, it would be misleading to speak of events that cause these experiences, as they might be there from the beginning, but rather about states of involuntary pain, loneliness, disease and so on. Might there be a kind of health that is compatible with experiences of such traumatic events or painful states? Might it even be possible to speak of suffering as a kind of health?

I have elsewhere argued that it is possible to distinguish a concept of health that we could call *existential health* which is not another dimension alongside mental, physical or social health.[9] Not even the "spiritual health" recently taken up by the WHO as a fourth dimension of health is the same thing as what I mean here by existential health. Instead of being another dimension of health, existential health, I would suggest, is an aspect of health encompassing all other dimensions of health; it is our own intentional reflexive relation to experiences of both health and illness. Existential health is a first-person perspective, then, and a first-person perspective in the sense that it is my own relation to infirmities and calamities, my own or others. This means, among other things, that it is possible to be ill and healthy at the same time. Existential health does not stand in contrast to a particular disease; I could very well be the victim of many ailments without lacking in existential health. Existential health is not about being free from all possible diseases or pains or struggles, but rather a question of how I relate to each and all of them. Thus it is an aspect of how our subjectivity is involved in our experiences of pain and misfortunes—in other words the solidarity between myself and such experience.

I would claim that existential health is related to suffering. I suggested above that suffering involves a transformation or transfiguration of pain such that suffering allows me both to encounter and confront pain. Suffering is an active passivity, but not only as a static mode of being. Suffering is rather a dynamic being in the world where we achieve health through learning how to suffer. If suffering is not a passive sensation but actually an active cultivation of a particular relation to the experiences of pain and misfortune, then suffering is a work that aims at existential health and in a sense already is that kind of health. It is the work of acknowledging a pain or misfortune, establishing a kind of solidarity, and learning how to endure it. Please note that I do not use the verb *accepting* but prefer speaking of *acknowledging* pain or misfortune. I am wary of using *accepting* as it might sound as if I would suggest that suffering means giving in to pain or misfortune. I do realize, of course, that there might be situations when acceptance is the only viable option we have in the face of chronic pain. But even then, perhaps, accepting might not be the same thing as giving in to pain; even accepting pain can mean that we establish a minimal distance between the pain and our existence. If we can say, for instance, that I accept that I might be in chronic pain but only in the sense that this is how it must be and not how it should be, then we gain this minimal distance that allows us a kind of freedom. This is what I mean by *acknowledging* pain or misfortune as somehow inevitable without giving in to them. This minimal distance is important in that it allows a relationship between suffering and freedom and even hope, something I will return to below.

If suffering is an active passivity that could be referred to as work or learning, how do we go about learning to suffer? There are many aspects of such work, and this is not the occasion, nor am I the person, to address all of them in detail. But from a phenomenological perspective there are some traits of such work that also shed some light on what kind of personhood is involved in the work of suffering. There is a question then about subjectivity, and what kind of subjectivity conforms to suffering or needs to be laid bare to understand the work of suffering. To begin with, and consistent with what I have been emphasizing all through this chapter, the work of suffering entails a first-person perspective. The work of suffering involves the acknowledgment that a particular pain or misfortune really has struck me. This is not, I think, a trivial observation. There might be obstacles to such an acknowledgment, such as the phenomenon of repression, either of a subjective kind in that I refuse to acknowledge that I actually suffer from pains and misfortunes, or of a more objective kind in the suppression of pain, misfortune and death in the social imagination of contemporary society. As Rita Charon puts it in her now classic *Narrative Medicine*:

> The price for a technologically sophisticated medicine seems to be impersonal, calculating treatment from revolving sets of specialists who, because they are consumed with the scientific elements in health care, seem divided from the ordinary human experiences that surround pain, suffering, and dying.[10]

This divide has a phenomenological correlate in the experience of the dys-appearing body, but without the re-appropriative movement of solidarity between me and my pain. The refusal to acknowledge pain and misfortune as integral aspects of human existence does not only give rise to an extrinsic relationship to those aspects and an objectification of one's own existence, but also is a serious obstacle for any processing of such experiences, making them mine, and thus to the work of suffering.

Further, as suffering is a dynamic mode of being in the world, learning to suffer is a process that takes time. Traditionally, the way one learns to suffer is through patience, and patience means enduring. On the one hand, patience means giving up something, most of all giving up time. Whatever patience is for, for instance recovery, patience means that what I yearn for will not take place now but later. Patience also means acknowledging a certain incapacity or impotence in one's self, not being able to achieve the longed for recovery, by one's own power. But on the other hand, in this passivity of patience, it is not the same thing as just waiting something out. Patience can also be learned through developing certain habits through actions and deeds that do not strive for immediate wish-fulfilment, including physical therapy and occupational training.

Those actions and deeds are such that they train us to live in an active anticipation of a desired goal we cannot master through our actions and deeds and have to receive as a gift, even as an anticipated gift. This also shows, I hope, how patience is not just a matter of a mental or intellectual attitude, but an embodied practice. It is, so to speak, the body's way of enduring the embodied alienation that is a result of pain and misfortune. Consequently, with regard to suffering, patience is not the same thing as suffering but can be practiced independent on suffering. The work of suffering needs to draw on patience in learning how to endure pain and misfortune.

Finally, since the work of suffering is a process, there is a narrative quality to the subjectivity presupposed by this work. The narrative quality of subjectivity should here be taken as a contrast to an atomistic, fragmented view of human existence currently touted as medicine by a bureaucratic health care. Charon again: "Sicknesses declare themselves over time, not in one visit to the consultant."[11] The work of suffering could be seen as a way of reconfiguring the relations not only to one self but also to others as well as the world. Some infirmities will make it impossible for me to do things that I have been used to doing, maybe things that have been important to me, and through the work of suffering I must find a way to take leave of them and find a new way of relating my existence to my environment and vice versa. This reconfiguration has a narrative quality, in that to make sense of my experience I must be able to tell what it was like before, what I now experience, and what I hope for will be the outcome of what has happened to me. This is not saying that it belongs to the work of suffering to write an autobiography, even if it could do. Narration, as I think of it here, is a much broader category than the literary, as it will do the work of re-appropriating the painful or traumatic experience within the realm of my existence. Narration is here a work of acknowledgment in the aforementioned sense, making sense of the discontinuities that painful or traumatic experiences entail by narrating them and thus making them continuous with my broader experience—without denying their discontinuity as such. Narration in relation to the work of suffering also explains how it is possible to bridge the gap between one's self and an other. Pain or misfortunes can indeed be lonely and alienating events or states, but through narrating them it is possible to re-establish a relation to one's significant others. Suffering is not as such a private or individual experience. It is through narration that we are able to give an account of relationships extended in time. If pain can be an alienating event or state that not only divides me from myself but also from friends, relatives and society, it is through narration that these relationships can be restored or even redeemed. The narrative quality of the work of suffering is thus instrumental in transforming the event of pain, gaining a relationship to it that is characterized by an active passivity.

Acknowledgment, patience and narration are all equally important in the work of suffering as a part of the transformation of pain. Talal Asad makes a similar point:

> What a subject experiences as painful, and how, are not simply mediated culturally and physically, *they are themselves modes of living a relationship*. The ability to live such relationships over time transforms pain from a passive experience into an active one, and thus defines one way of living sanely in the world.[12]

The ability that Asad is talking about is what I would call suffering, the transformation the work of suffering and a "way of living sanely in the world" of existential health. Existential health means gaining a relationship to the world, parts of which overwhelm me in ways that could be described as violent in the sense that they intrude themselves on me. I have perhaps never asked pain to arrive, at least not intentionally so (excepting that I might be in denial of my real wishes), but nevertheless I have to deal with it once it has established itself here. Suffering, the work of suffering, is the means I have to affirm the solidarity between me and the pain, to work through the gap between being a body and having a body within my embodied life.

Conclusion

Only vulnerable creatures suffer. Such an assertion sounds obvious, of course, but I use this pleonasm anyhow to emphasize the centrality of suffering for human agency. To not want to suffer does not only mean engaging in dreams of invulnerability that could also have detrimental effects on the possibilities of engaging with pain and misfortune from the side of society. Not recognizing the vulnerability of our human creatureliness would of course be one thing if it is deliberately chosen by an individual. But if invulnerability and the dream of being in control becomes a part of the social imaginary—and I suppose that it is not too controversial to suggest that this is to a large extent the case in our present Western society—then such ideas of human existence will start affecting health care, social securities, the school system, attitudes toward refugees, as well as the fabric of relationships that make up the associations, neighborhoods and families of civil society. If suffering is an essential characteristic of human existence, then its active passivity needs to be recognized as such to let the work of suffering take its course with the help of institutions that support it.

But the recognition of suffering is not only important because of the effects it might have on the social imaginary. It is also important on the existential level for us as persons. The suffering of pain has been the focus of most of this chapter, but I also mentioned suffering love, the experience of being overwhelmed by a passion. In a fuller account of suffering than mine, it would be imperative to also give an account of how desire and love affect our mode of being in the world; how, perhaps, only vulnerable creatures love. Vulnerability, the human characteristic that is the condition of possibility of being able to suffer, has to do with the ability to receive something from outside of our own selves without making preconditions of how reception should be achieved. Vulnerability is then the ability to be surprised, not only by pain but also by joy, happiness, pleasure. That is, perhaps, another dimension of the work of suffering.

Notes

1. This does not mean I regard the distinction between different modes of phenomenology, or between Martin Heidegger, Maurice Merleau-Ponty or Paul Ricoeur, as radical. For more on this, see my *Heavenly Bodies: Incarnation, the Gaze, and Embodiment in Christian Theology*, trans. Carl Olsen (Grand Rapids: Eerdmans, 2016), 19–30.
2. Judith Perkins, *The Suffering Self: Pain and Narrative Representation in the Early Christian Era* (London and New York: Routledge, 1995), 88f.
3. Ibid., 118.
4. For further thoughts on suffering, some of which have inspired my own account, see Donna M. Orange, *The Suffering Stranger: Hermeneutics for Everyday Clinical Practice* (New York and London: Routledge, 2011).
5. Maurice Merleau-Ponty, *Phenomenology of Perception*, trans. Colin Smith (London: Routledge, 1992), 82.
6. Drew Leder, *The Absent Body* (Chicago and London: The University of Chicago Press, 1990).
7. Such an account seemingly could imply that the individual's experience of a healthy state is normal and prior to experiencing pain and/or sickness, which certainly is not the case for everyone. This is not my point here, however, nor do I think it is Leder's. But even if the individual's experience of pain is more original than the experience of health, I assume that such a distinction still could be made, where health is the horizon of pain (and/or perhaps vice versa?).
8. Jacques Lacan, *On Feminine Sexuality: The Limits of Love and Knowledge, Book XX: Encore, 1972–1973*, trans. Bruce Fink (New York and London: W. W. Norton and Company, 1999), 119.
9. Cf. Ola Sigurdson, "Existential Health: Philosophical and Historical Perspectives," *LIR Journal* 6 (2016): 7–23.
10. Rita Charon, *Narrative Medicine: Honoring the Stories of Illness* (Oxford and New York: Oxford University Press, 2006), 6. Cf. Jeffrey P. Bishop, *The Anticipatory Corpse: Medicine, Power, and the Care of the Dying* (Notre Dame: University of Notre Dame Press, 2011), 20; Ola Sigurdson, "The Body of Illness: Narrativity, Embodiment and Relationality in Doctoring and Nursing," English translation of "Sjukdomens kropp: Narrativitet, kroppslighet och relationalitet i medicinsk praktik och omvårdnad," *Kritisk forum for praktisk teologi* 31, no. 123 (2011): 6–22, http://hdl.handle.net/2077/38639
11. Charon, *Narrative Medicine*, 7.
12. Talal Asad, *Formations of the Secular: Christianity, Islam, Modernity* (Stanford, CA: Stanford University Press, 2003), 84. Cf. Sigurdson, *Heavenly Bodies*, 534–7.

6 The Living Body Beyond Scientific Certainty

Brokenness, Uncanniness, Affectedness

Thor Eirik Eriksen

Introduction

Living humans with living bodies are part of this given universe of meaning we call life. Inevitably, a *living* unfolding of life also entails suffering, annoyance, disturbances, imbalances, or what we can describe as "brokenness" or "experiences of brokenness." Such events are linked to meaning in a basic sense, and through these we are quite often thrown back on ourselves, our lives and our place in it all. Further, these are events where the bodily sensibility fuses with thinking, wonder, interpretation, and often the wish to share these with someone. Consequently, in such processes we rely heavily on language. From a phenomenological perspective, it is therefore essential to provide and strengthen a language that describes the *living* aspects of brokenness and suffering. This is especially important in an era in which a medical-scientific language of pathology, including static descriptions of symptoms and diagnoses, represents both a lifesaving and troublesome reference point in people's lives.

We can regard this scientific production in medicine as serviceable, since it attests to medical breakthroughs and provides solutions to serious health problems. Furthermore, it is an expression of a well-motivated search for knowledge on behalf of health problems still awaiting answers. However, we need to reflect on how this production *can* also have a masking effect. The process of constantly providing new pieces of the puzzle of the object-body (*Körper*) risks obscuring the non-objectifiable, experienced, living body (*Leib*). In the face of this challenge, we need to engage in critical thinking which may enable us to decipher all those fine nuances relating to scientific progress in the field of medicine. The discourses relating to such matters involve profound ontological and epistemological issues, among which are questions concerning the boundary issues between normality and pathology. It is, however, beyond the scope of this text to elucidate the full complexity of this disputed field. Consequently, my starting point will be a narrower dimension of the problem such as the simplified distinction between *object* and *phenomenon*. More precisely, I suggest that an underlying and problematic aspect relating to a partially *unreflected* medical-scientific expansion resonates with Jean-Luc Marion's general description of an *asymmetry* in the object-phenomenon relationship.[1] The object, provisionally determined as the thing put before us or "the thing external to the thinking mind or subject" is here supposed to *precede* the more unclear, "subjective" and demanding phenomenon which manifests itself.[2] The object takes precedence, and, as Marion reminds us, it "takes up space and attention, while the other phenomena hide in the background, pushed aside by the object."[3]

In the *first* section I will elaborate on this dichotomous characteristic of objects and phenomena and the idea of asymmetry, taking phenomenology as a reference point for both. Contributors in this tradition such as Marcel, Heidegger, and Marion can be used to challenge certain kinds of object logic and support an attempt to upgrade the phenomenon's status vis-à-vis the object. *Upgrading* here refers to an attempt to highlight different nuances and qualities related to rich descriptions of phenomena, for example a phenomenon such as tiredness. The aim will be to open up this phenomenon without being committed to what we might describe as psychological and medical object-descriptions of, for example, burnout and chronic fatigue syndrome (CFS) which are presented to us in diagnostic manuals. Consequently, this is not a crusade aimed at undermining a legitimate and mainly biologically founded object-approach in medicine, but instead a reminder of the possible limitations of this reasoning and the emphasizing of other ways to explicate and describe human suffering.

The second section explores the phenomena of brokenness, uncanniness and affectedness. These terms will be regarded here as markers and pointers marking out a region or a field to work with—and dwell within. Common to all of them is that they point to the living human's conditions of existence. The task here is to put such markers into play in a way that can safeguard their flexibility, indefiniteness and dynamic disposition. By treating them as open and flexible concepts, we can also pave the way for, or strengthen our receptiveness of, complex, rich or saturated phenomena. This flexibility, however, is not introduced in order to beautify pain and suffering. Instead it concerns the necessity of providing access to dimensions of brokenness that are not accounted for through fixed concepts, diagnostic classifications and scientific evidence.

Objects, Phenomena and Graded Phenomena

It is obvious that we cannot escape our need to *order* our existence. There is a fundamental and natural human yearning to decipher, sort and classify what is foreign or unknown, friend or foe. This also includes all the ordinary daily activities where we plan, organize, arrange and systematize our everyday existence. Consequently, creating order from disorder satisfies a basic need and safeguards our existence. When disorder and confusion become *disorder* in the medical sense of this word, the medical research community legitimately tries to create or re-establish order by producing causal explanations, accurate symptom descriptions, defined diagnoses and systematically proven interventions. However, one must strive to clarify when this desire to order contributes to the fulfillment of undeniable human needs, and when it culminates in controversial and possibly counterproductive outcomes. An example of the latter is discussions concerning medically unexplained symptoms (MUS), and contested diagnoses such as grief, gaming disorder and ADHD.[4] More generally, it refers to the outcomes of a scientific search for positive certainty where one strives to *transform* quite complex phenomena into medically intelligible objects, including the transformation of complex and life-infringing experiences of brokenness.

A broad discussion related to the diversity of possible outcomes of such a process, i.e. to what extent the transformations deliver "pure" or constructed, strong or weak objects according to a set of scientific standards, is however beyond the scope of this investigation. This is also the case for the possible differences between the "objects" of medical science and the shape these take in a completely different context, i.e. clinical practice. For now, it is sufficient to state how the object, in all its different forms, plays an important role in the ordering efforts in medical research. What then is the object's evident advantage in all this? According to Marion, the object is always preferred because, contrary to the phenomenon, it can be pinned down, i.e. quantified according to set parameters, predicted, reproduced, and attributed to specific causes.[5] Further, and of special importance in the field of medicine, the quality assurance of the object is a necessary condition for *producing* clear, distinct and reliable knowledge. This is also what ultimately contributes to reliable actions and interventions in medical practice. For example, when using a comprehensive arsenal of medical instruments to accurately determine the object identity of a broken bone, this approach is optimal. In contrast, when a doctor is faced with a weary and exhausted patient, the absence of objective characteristics may be the only thing that can be objectified. The fatigue eludes quantification, is difficult to predict, and protests against being subject to a causal explanation. Even after consultation and examination, neither the helper nor the patient manages to pin down or constitute this event, i.e. the patient's experience of being fatigued.

Apparently, such an event and the related symptoms do not *qualify* as a detectable object amenable to advanced and well-proven medical tools. The fully legitimate and necessary task of trying to *transform* a complex and rich experience into a kind of tangible medical entity thus appears unsuccessful. This suggests that the constructive part of this transformative process does not produce a medically intelligible result. The medical professional is instead left with a comprehensive patient narrative and an incoherent collection of symptoms (a symptom-syndrome). A tendency to postpone further examination or referring the patient to other health professionals could then be the possible endpoint. The reason could be that this phenomenon, in this context and compared to the object, represents an *unreliable* source—at least, as long as the ambition is to reach objective certainty. The qualified, preferred object and the often disqualified phenomenon form the basis for what Marion terms asymmetry: "According to its mode of knowledge, its *ratio cognoscendi*, the phenomenon of the object type prevails indisputably over the nonobjective phenomenon, which it devalues as uncertain, imprecise, and confused—in short, as being at the margins of knowledge and quasi-irrational: subjective."[6] Consequently, the phenomenon refusing to fall into line with our attempts to define it and pin it down; in reality, it appears simply as a failed object, a residual category. It does not fit into the knowledge-gathering ambitions and is afforded no status in relation to explaining the case of the human body.

A first step in attempting to pave the way for the phenomenon and raise its status will deal with reflections on human suffering, which, as far as possible, are unfettered or detached from a dominant medical terminology and discourse. As mentioned earlier, the intention is not to devaluate or degrade diagnostic classifications, but to enter a reflection on phenomena which is not bound or constrained by the systematizing efforts in medicine. *The consequences, however, are not that the object is rejected, but that instead it is radically incorporated into and subsumed by the phenomenal.* This means that here, in a purely philosophical exploration, we choose to regard *all* the events that are in different degrees manifested in a human life as phenomena. *The consequences of this are that objects and non-objects are incorporated into a common phenomenal point of origin—a univocal phenomenality* (my italics).[7] The objects here retain their undisputed qualities as identifiable, delimitable, measurable, and open to manipulation, while within a framework of phenomenality they are considered decimated, shrunken or diminished phenomena. In this way, an established asymmetry is challenged because

> the nonobjective phenomenon unquestionably wins out if its mode of being, its *ratio essendi*, is taken into account. For since it does not depend on the understanding to constitute it, but arises from itself, without any warning to prepare us, and without any repetition to accustom us to it, it imposes itself as an actuality without cause, autonomous, spontaneous, fully realized of itself and always in advance of any knowledge we might later glean from it.[8]

Consequently, this inability to immediately grasp the phenomenon is not a problem; on the contrary, it is the premise that attests to the phenomenon's ability to maintain itself and its independent status. This delay in the constitution of understanding does not need to be a problem either;

rather it is a condition for managing to establish a living relationship with the phenomenon that announces its arrival. Similarly, a lack of causal understanding need not merely be a problem, since it can also be an invitation to develop our receptiveness to phenomena. That is, the "why" question, after the failure to provide an adequate, medical answer to it, can temporarily be put on hold in a process which for some people provides an opportunity to situate themselves in the lifeworld.

The establishment of this univocal phenomenality is a preparation for the next step, taking us further beyond the provisionally postulated dichotomy between object and phenomenon, i.e. between supposed mutually exclusive extremes like "pure" objects on the one hand and saturated phenomena on the other. It is assumed instead that objects and non-objects are spread out along a continuum. The phenomenon is therefore assumed to have different points of impact along an infinite continuum, and to have different degrees of phenomenal charge, complexity, configuration, fullness, richness, mystery, or saturation.[9] This implies that the fundamental phenomenological concepts we apply in our descriptions should clear the path into a phenomenal world that allows complexity, diversity and richness.[10] The terms that are unfolded in the following—brokenness, uncanniness and affectedness—are intended to be phenomenon markers that can guide us on this journey.

Brokenness

In our everyday life, we usually move ourselves from one activity to another without paying close attention to all the facets of our moves and actions in different environments. The acting, feeling, sensing and living body leads us to what we usually experience as a kind of unbroken unfolding of life. This body, which is indisputably mine, unfolds in a continuous collaboration with other human bodies and the world. What is meaningful in this context is that these dynamic and lively aspects do not become less crucial when the body protests, obstructs, and in various ways puts obstacles in our way. That is, when *brokenness* occurs, which in some way marks a transition or a shift. Etymology gives some indications that this transition can be intrusive and dramatic. The adjective "broken" refers to something that is "separated by force into parts, not integral or entire." The noun "break" alerts us to the "act of breaking, forcible disruption or separation" and marks a "sudden, marked transition from one course, place, or state to another."[11]

Such experiences of brokenness, to the extent it is recognized that they concern complex and saturated phenomena, can consequently appear in a variety of formats and modes of expression. A heavy lift, followed by a prolapsed disc and intense prolonged back pain, may be an example of such a "sudden, marked transition." This is seemingly an experience of brokenness that is understandable, i.e. one in which the causal relationship is clear and where the pain can be fairly precisely localized. However, even this intelligible experience of brokenness can impinge upon my entire unfolding of life, restrict my mobility, affect my mood, change my focus, disrupt my sleep and even interfere with my way of relating to others. Consequently, through an incident of well-defined and delimited brokenness, a variety of meaning-bearing and meaning-giving expressions of life such as fatigue, tiredness, unwellness, powerlessness, sadness, anxiety or despair can manifest themselves. Such expressions are deeply rooted in collective human nature and, in reality, represent conditions that permeate our existence and our humanity.

A broken relationship or marriage can apparently represent a very different brokenness event, characterized by strong emotions, conflict, fear of getting hurt, and loneliness. However, there is nothing to suggest that this experience can be isolated as something "psychological." That is, as something "in the mind" concerning purely cognitive, mental and/or emotional disturbances. The experience can be pervasive and saturate our bodily inner being, where stomach pains, anxiety, muscular pain, headaches, and lack of energy can unfold in an embodied totality: a totality that is uninformed and indifferent as regards the borders between psyche and soma. What is therefore common to the broken back and the broken marriage is that we are in contact with phenomena whose extent, complexity, and phenomenal fullness cannot easily be captured. Should we nevertheless insist on intervening, including in a clinical or therapeutic setting, it is self-evident that the complexity and richness of such experiences must be reduced or limited. The clinical and therapeutic tools available can potentially support a process aiming at clarity, overview, stability and order. In this setting, and in the search for object characteristics, brokenness events are transformed into clinical problems that are subject to problem investigation.

What does it mean, therefore, for "something" to be identified as a problem, and what will this mean for our understanding of experiences of brokenness? One philosopher in particular who has studied this is Gabriel Marcel.[12] His point of departure is that the problematic, i.e. a problem, is something that can be set out, classified, analyzed and solved. A problem is a question posed concerning something manageable and solvable.[13] This solvability has a strong appeal and in this world of problems there is thus, not unexpectedly, "an infinity of problems, since the causes are not known to us in detail and thus leave room for unlimited research."[14] It is precisely this unlimited potential that gives us, and especially the sciences, carte blanche to constantly take new steps in a problem investigation. However, this does not take account of the fact that humans find themselves in a *situation* that exceeds the scope of the problem perspective and cannot itself be made into an object. Experiences of brokenness transcend problems, and they concern that which, according to Marcel, constitutes the problem's natural contrast—*a mystery*.

A mystery in this context refers to that which is characterized as not being a problem. This means "something" that resists or does not allow itself to be subjugated—"to an exhaustive analysis bearing on the data of experience and aiming to reduce them step by step to elements increasingly devoid of intrinsic or significant value."[15] The reception and incorporation of the mystery requires a willingness to refrain from a one-sidedness that turns the mystery into a stripped, decimated, and decoded problem (or object). An important point is the recognition that I *am* my body, as opposed to possessing it and managing its instrumental functions. Marcel argues that this recognition is only possible "in so far as for me the body is an essentially *mysterious type of reality*, irreducible to those determinate formulae to which it would be reducible if it could be considered merely as an object."[16] As a mystery, our embodiment is not, therefore, a *thing* we can determine scientifically and subject to technical mastery or analysis according to a problem formula, without stripping it of intrinsic value and meaning. If we are to follow Marcel's train of thought, any attempt initiated by ourselves or others to take on the role of investigator and analyze the body as a problem unit could easily end up with us setting aside ourselves as existing. It is worth noting that discussions about the content of modern patient care repeatedly return to this particular challenge. Once again, this clearly does not obviate the need for investigations of health-related *problems*, but instead reminds us that the broken body houses both problems and seemingly ineradicable mysteries.

How then should we pursue our search deeper into the labyrinth of brokenness and the human experiences of it beyond the descriptions of problems? One possibility may be to provide some space for a phenomenon that can serve as the messenger for the powerful, importunate, and relentless, cf. the previous etymological considerations, namely pain. The simplest description of this phenomenon can be accommodated in this single assertion: pain hurts. It has something about it that makes me unable to be indifferent to it. It bothers me, and it speaks to me in the form of a hushed or pronounced whisper, and even a deafening prosecution. It is experienced, suffered, and lived through a characteristic and irreducible subjectivity. It is *my* pain, my pain narrative, my pain sensitivity, and my pain tolerance that are played out within my lifeworld horizon. Clarifying what it signals, where it originates, and how I should relate to it is a challenge. As a phenomenon, it is saturated and charged and concerns appearances that relate to meaning.

Despite its obvious and intrusive signal-informative or communicative character, pain is just as much a phenomenon that is difficult to share verbally with the public world. It has an inner aspect that is difficult to put into words. For this reason, experiencing pain can also be perceived as a heavy, costly, and lonely burden to bear.[17] It is obviously unlikely that attempts to translate pain into appropriate medical data will curb such feelings.[18] A worthwhile pain-relieving clinical intervention may have such an effect, but the objectified pain that remains as a residue when a meaning-saturated and complex pain event has been captured and confined will always represent that which follows, that which is secondary and only a limited part of the pain event that unfolds. As a phenomenon, pain therefore shatters the framework for many of our attempts to pin down, define, monitor, and intervene. As an aspect or dimension of the broken body, pain remains an embodied and embedded mystery—or a "meta-problematical" challenge. That is, in Marcelian terms, "something in which I find myself caught up,"[19] which "cannot be reduced to detail"[20] and which "transcends every conceivable technique."[21] However, and thanks to medical discoveries and the access to (sometimes) effective pharmaceutical remedies, it is possible "to degrade a mystery so as to turn it into a problem" and initiate pain-relieving measures. Nevertheless, the challenge that remains is to realize or understand *when* this degradation, i.e. the transformation from non-object to object, is necessary and when it could potentially lead us astray and in itself become a pain-inducing process.

Uncanniness

In the previous section, we have seen that brokenness events can be abrupt, vigorous and painful, but they may also entail more indistinct and non-identifiable transitions. Uncanniness can, for the time being, serve as a marker for events that relate to the latter keywords. As an event, it may concern more indeterminable and unpleasant experiences by not being in place, not having an adequate footing, not having the necessary contact with our self or others, or feeling uncomfortable with the place or position we have in our existence. In my meetings with exhausted individuals suffering from pain, stories are often told that relate to this. Sometimes they describe what it is like to feel maladjusted, out of tune and like a stranger in their own life. It has to do with not being in harmony with either themselves or their surroundings. It concerns the lack of fixed reference points, the recognizable, the familiar and being "in flow." Thus, it is more than the physical body that revolts or thoughts that lead these individuals astray.

In an extension of such descriptions, it will be natural to involve Heidegger and his discussion of this phenomenon in *Being and Time* (2010).[22] Uncanniness, or the more direct translation "unhomelikeness," can be traced back to the German *Unheimlichkeit*. This state of not-being-at-home is first evoked in Heidegger's discussion of anxiety. For Heidegger's "Dasein," which is Heidegger's term for a human being, uncanniness represents a basic mood [*Grundstimmung*]. In a passage of the book, this term is applied as follows: "Entangled flight *into* the being-at-home of publicness is flight from not-being-at-home, that is, *from* the uncanniness which lies in Dasein as thrown, as being in-the-world entrusted to itself in its being."[23] The last two clauses relate in short to how we are thrown into the world and into existence. More importantly, this points to the situation that unhomelikeness already, in the thrownness, conditions Dasein's existence. Consequently, unhomelikeness is not just something that only occurs as an unfamiliar and transient brokenness in relation to the homely and familiar. Heidegger elaborates: "Tranquilized, familiar being-in-the-world is a mode of the uncanniness of Dasein, not the other way around. Not-being-at-home must be conceived existentially as the more primordial phenomenon."[24] Uncanniness is therefore a fundamental mode that ensues from being *there*, as the *place* or the aperture where being can make sense. The first clause of the quote points out that we try to keep this unhomelikeness at a distance, through an *"entangled flight into the being-at-home of publicness."* We flee into "the They," the public "anyone" and into idle talk, ambiguity, and curiosity. Heidegger describes this movement of losing oneself as fallenness, and it refers to how we can be swallowed up or absorbed by "the world." The role or task of uncanniness is to put this propensity to flee and lose one's self under pressure: "uncanniness pursues Dasein and threatens its self-forgetful lostness."[25]

Upon this stranger's arrival, human beings are thus exposed to pressure and are challenged. Almost imperceptibly we are no longer encapsulated in the safe and familiar, but this feeling of unfamiliarity nevertheless cannot be determined qua content or form. Likewise, we are reminded that something is not as before, but not such that this *before* versus *here and now* can necessarily be sorted or organized. We may feel threatened or exposed, but without being able to identify the threat. Furthermore, and despite the fact that this irrefutably concerns *my* existence, this unbroken proximity to the stage where it is all played out gives no benefits in relation to bringing the discomfort to clarity. With a view to our ability to grasp uncanniness as a thing or object, or force it into a definitional strait jacket, Withy confirms what has already been suggested earlier, namely that "the concept of the uncanny is itself uncanny."[26] This means that it cannot be mastered or domesticated. This leads me to my main concern, namely that our attempts to encapsulate and reduce this phenomenon only serve to wrench it from its phenomenon status. In a recasting where we believe to have taken possession of this phenomenon and transformed it into something familiar, recognizable and delimitable, all contact is thus also immediately lost. Uncanniness is therefore a standing invitation that requires both a distinctive inquiring attitude and a non-dogmatic stance on what an answer should be. To the extent that an answer is possible at all, the answer according to Withy is very simple: uncanniness. "It is because the question that we have raised is the one that makes us uncanny that the question of the uncanny is most properly worthy of question."[27]

In the foregoing and via the unhomelike and the unfamiliar that put human self-forgetfulness under pressure, the fundamental moods of uncanniness and anxiety are inextricably linked. How then can the anxiety be described, assuming that this phenomenon cannot be reduced to any great extent to merely an object-phenomenon? Within the scope of this text, it is not possible to go into this in depth, so I will therefore, in line with the general theme of this chapter, emphasize the complex nature of anxiety—specifically, that it simultaneously involves withdrawal and revelation. With regard to the first aspect, Heidegger reminds us that the importance of our world and what we surround ourselves with, including the objectively present, changes character in the encounter with the experience of anxiety. In the anxiety, the world has nothing more to tell us: "The world in which I exist has sunk into insignificance."[28] That means that the anxiety closes and narrows. It makes my perspective limited and my world smaller. However, given the overall theme of this chapter, the second aspect of anxiety—*revelation*—is of particular significance. Heidegger is very clear about this point, stating, "Anxiety *reveals* in Dasein its being toward its ownmost potentiality for being."[29] The anxiety here is that which discloses, reveals, and displays. It not only reminds me of but also discloses myself in my thrownness and reveals the world in its worldhood—"that is, as a meaningful context in which we dwell."[30] It induces how "I am individuated as a case of Dasein and so as a sense-maker."[31] How, then, can we understand this simultaneous withdrawal and revelation in the anxiety?

One possibility is to thematize our relationship with the Other, and especially a phenomenon such as friendship. A particular challenge that can manifest in anxiety is that I may come to place myself, for myself, as insignificant and unfamiliar in relation to others. In this, I ascribe to myself, almost imperceptibly, an evaluation that tones down, mutes and even places me on the sidelines or outside of what, for me, constitutes my human communality. I can no doubt try to compensate for this feeling of self-shrinkage by accelerating and intensifying my zealousness and need for activity in the world. The outcome of this self-initiated flight only serves to exacerbate the anxiety and reinforce the encapsulation. The paradox of this self-shrinking movement is that my fellow humans, every possible traveling companion on life's path, are pulled along in the undercurrent. The significance they have, or have ever had, is no longer clear to me. The collegiate community that has previously enveloped my working activity is now only a collection of single individuals who are there separately and for themselves. Those closest now are situated the farthest away. Even family is transformed into the unfamiliar. The experience of anxiety is total and pervasive. It anticipates and precedes any schematizing according to thought, feeling, and action. Anxiety is felt and lived. It is, more than anything else, distance.

The paradoxical and enigmatic element to this is that the shrinking and restrictive movement almost simultaneously involves disclosures. This can be linked to the obvious, that anxiety, in the withdrawal, puts something into play, or more precisely, puts something at stake. Self-understanding in this is under review and revision and brings me to questions like these: "What is it like to find one's self in the context, in the community, in the world?" "What, or who, or for whom is it that I am supposed to be in all this?" "What is involved in this possibility or impossibility?" Just as imperceptibly as withdrawal's shrinking movement becomes known, anxiety enables a rediscovery of context or, more precisely, meaning connection. I may rediscover myself as who I am or have to be—*there*. This relates to a return from self-forgetfulness and the slow-growing feeling of indifference. Relationships now emerge in a new light. They reappear, as they have done countless times before, through this uncomfortable and coercive reorientation.

We have previously mentioned how objects and phenomena are not separate entities that pertain to separate realities, but projections that are derived from univocal phenomenality and can have different impacts on an infinite continuum. This is also the case with withdrawal and revelation. According to Withy they are "not separate events but two ways of describing the same thing. Self and world are likewise not opposed but are two ways of describing the same thing: Dasein's being in the world."[32] This vast span in the experience of anxiety attests to its being a phenomenon that cannot be objectified and decimated, without paying a price. To subjugate anxiety to the scientific method means, as Heidegger reminds us in *Zollikoner Seminars* and later, to insist that it is an object among objects of conceptual representation.[33] With an eye to anxiety, Heidegger therefore states the following: "Anxiety and fear are not objects. At most, I can make them a *theme*."[34] As a theme, anxiety is more than merely an illness; it is a window to life and says something crucial about what and how a human is.

Affectedness

When humans are faced with challenging brokenness events and experience pain and anxiety, one question often recurs: why? Accompanying questions are sometimes the following: "What has caused it?" "What has triggered this?" "What have I done?" This quest for explanations, given our desire for order, can lead the afflicted individuals into a very complicated web of causal hypotheses. Fortunately, such quests will occasionally provide useful answers. Irrespectively, the basic idea, regardless of whether one finds an answer or not, is that something can be identified as *the* source of my afflictions. Being *affected* will consequently imply in such a context that one is exposed to a bunch of active causes (see

Aristotle's *causa efficiens*). By linking the thread back to the introduction, we can assume in simple terms that such causal reasoning requires an ability to identify some object dimensions. This means we are dependent on identifying and isolating potential causal factors just as strongly as their possible manifestations in object format (e.g. biologically detectable abnormalities or diagnosis labels).

From a phenomenological point of view, however, this causal reasoning is a call for thinking and can (and should) be reflected on. Thus, being affected, or affectedness, can be a starting point for rethinking the concept of causality. In this regard, Jean-Luc Marion is one of those who strongly argue that this causal reasoning has not only a disclosing but also a masking function. He insists, therefore, on the necessity that "we open the question of givenness itself, according to its own fold, without undue reference to efficient causality."[35] Among other things, Marion's insistence relates to how causality does not allow phenomena to be phenomena:

> I suggest that phenomena as such, namely as given, not only do not satisfy this demand [the demand for a cause], but far from paying for their refusal with their unintelligibility, appear and let them be understood all the better as they slip from the sway of cause and the status of effect. The less they let themselves be inscribed in causality, the more they show themselves and render them intelligible as such. Such phenomena are named events.[36]

Affectedness has precisely this event characteristic.

This gateway to the given nature of phenomena also casts light on the reason for Marion's understanding of the body, which he refers to as "the flesh."[37] For Marion, this living body is a stage upon which certain given, saturated phenomena play out, i.e. a phenomenon "touches me," "reaches me, " "arrives to me" or "falls upon me."[38] This means that I, through my living body, find myself caught in a stream of phenomena that arrive, come into contact with me, and affect me. There are no exceptions here: "No phenomenon can appear without coming upon me, arriving to me, *affecting me* as an event that modifies my field."[39] This affectedness must here be understood uniquely in relation to Marion's philosophical project. His point of departure is that flesh affects itself, unconditionally. What is crucial here, as compared to the above approaches on causal thinking, is that the surplus of intuition, the excess and the overwhelmedness, affects the flesh and "saturates the horizon to the point that there is no longer any *relation* that refers it to another *object*."[40] Marion repeats: "The affection refers to no object."[41] This self-affection implies that we are again confronted with this *mineness* that we cannot escape from or pass on to others. The pain is my pain. The flesh shows itself to me, surrenders itself, as itself and from itself. Marion summarizes this entire train of thought as follows:

The flesh auto-affects itself in agony, suffering, and grief, as well as in desire, feeling, or orgasm. There is no sense in asking if these affects come to it from the body, the mind, or the Other, since originally it always auto-affects itself first in and by itself. Therefore, joy, pain, the evidence of love, or the living remembrance (Proust), but also the call of consciousness as anxiety in the face of nothing (Heidegger), fear and trembling (Kierkegaard), in short, the numen in general (provided that one assigns it no transcendence), all arise from the flesh and its own immanence.[42]

Here we come to the heart of the matter. This stated affectedness, or auto-affection, tells us nothing other than the obvious, that it is through the body, the flesh, that we suffer and live. What I am is what affects me—my flesh. It is worth noting here, in this absolute mineness, that Marion states, "it always auto-affects itself *first* in and by itself."[43] First, this means that we first live through this bodily mineness and its immanence. It means that only subsequently are we capable of responding, of directing our attention, of grasping and comprehending possible relationships, including causal relationships.

My engagement in this subject is most closely concerned with precisely this, the subsequent, but not in the form of a process that reduces, weakens, and decimates the phenomenon. Instead it is possible to pave the way for a hermeneutic whose function is not to operate on objects or to ascribe their true meaning to given phenomena.[44] Hermeneutics has a far more important task, which is to "practice" ascribing meaning that redeems and releases the phenomenon in its givenness and strengthens the phenomenon's phenomenality. Marion notes here that "the more the phenomenon in terms of the flesh is saturated with intuition, the more it requires, to be visible to others, the word; sorrow, of course, but even enjoyment is not an exception."[45] The receiver can consequently challenge "even the absolute immediacy of a given saturated in terms of the flesh."[46] Self-affection or auto-affection can now become a springboard for fellowship with others, bringing us back to the starting point and to those who are afflicted and suffering. If we are to support a dialogue that preserves the living body's meaning-bearing and meaning-giving expression, we need words and concepts that do not, from the outset, require us to be confronted with something that can be captured in object format.[47]

There is one event that befalls humans that, more than many other events, protests vigorously against being subject to objectivizing strategies and causal explanations. Considerable attention is devoted to it in the medical academic community, the media and among the general public. It is the matter one refers to as fatigue, chronic fatigue or chronic fatigue syndrome/ME (CFS/ME). Consequently, and precisely because the medical term *fatigue* is a turning point in this discourse, I will instead refer to the phenomenon of *tiredness*. This is an exemplary saturated phenomenon that arrives without the affected receiver's or the helper's consciousness being able to respond with a satisfactory and intentional object-constitution. Through intuition, this phenomenon inundates and floods our targeted ability to resist. However, what matter does this concern? In replying, I will refer to a very comprehensive case description which lifts the fact of being tired out of any established medical conceptual framework. In the *Dictionary of the Danish Language* for the period 1700–1950, the noun *tiredness* is described as follows:

> Tiredness: who due to exertion of a corporal or spiritual kind or because of illness etc. (especially: of a transient nature) has lost a considerable part of one's powers, one's energy, vigor, or similar, and for this reason feels some lassitude and urge to rest; also indirectly: taxed by worries, life's tribulations etc. Furthermore who feels no joy, inducement, urge or need to continue or persevere in something; who does not (any longer) perceive any joy or satisfaction in something (that has been enjoyed to excess, that has been repeated too frequently, that has caused inconvenience, regret, or whose modest value one has recognized, etc.).[48]

This description manages to convey the simplest and the closest meaning, namely that tiredness tugs at the roots of our entire existential situation. It plays out in a life context and interplays with life events and life practice. Tiredness is also a phenomenon which we can assume to have different points of impact on our lives along an infinite continuum, and consequently to have different degrees of phenomenal charge, complexity, configuration, richness or saturation. Obviously, the tiredness that is elaborated on in this context does not primarily address the transient tiredness that manifests itself after a workout, a hard day's work or a late-night party. Instead, it concerns the variants of tiredness which is not transient and easily linked to obvious causes, and that to a greater or lesser extent have an impact on our lives. This also implies that tiredness can be regarded as a phenomenon which challenges and *questions*. It possesses the capability of questioning the manner in which we lead our lives. In this regard, and to the extent that this interrogation has a self-caring and self-protecting potential, it may be regarded as a statement for the defense from the depths of our humanity. In my objectivizing endeavors, it is easy to disregard the fact that I am thus being prosecuted. It is easy to forget that in striving to capture the tiredness, I may come to lose my empathy—and contact—with it. However, my body does not allow itself to be affected by these endeavors. The tiredness protests independently. It is autonomous. It manifests itself without regard for clarity of cause. It plays out on the stage that is my body with no regard for the body-host's analytical officiousness. If, over time, I should attempt to disregard or distance myself from its arrival, then the closing scene will consist in my finally having not merely to "lay myself down," but to be "shut down." This is the body's "no." I have no choice but to listen, despite my interpretive schemata falling short. This means that I can neither conclusively nor finally grasp such a phenomenon through a particular set of considerations because it bears a wealth of events that no single sweeping glance can capture. The event must be understood as absolutely unique, and also as what takes place before the event is deciphered and interpreted into an objectivizing schema.

Conclusion

Trying to bypass a way of thinking that fixes, trims and decimates, in this chapter I have highlighted some phenomena that may challenge our thinking about human beings and the question concerning what is human in humans. Brokenness, uncanniness, affectedness and associated phenomena such as anxiety, fatigue, and pain *all* represent demanding events that intrude upon this essential humanness. As I have emphasized in the foregoing sections, the preferred names for these events—brokenness, uncanniness and affectedness—represent important pointers and markers. Finally, and more precisely, they could be seen as "formal indicators" and "markings-out." As pointers or *formal indicators*, the latter term borrowed from Heidegger, they indicate a *direction*.[49] That is, they invite us to direct our attention toward aspects of life without being captured in a theoretical, distanced and objectifying posture. Dahlstrom reminds us that Heidegger's use of the notion "makes clear that he regards the 'formal indication' as a revisable way of pointing to some phenomenon, fixing its preliminary sense and the corresponding manner of unpacking it, while at the same time deflecting any 'uncritical lapse' into some specific conception that would foreclose pursuit of 'a genuine sense' of the phenomenon."[50] Consequently, the formal indications are starting points or beginnings that prepare the ground for a challenging journey of exploration into the lives of human beings.

As for the aforementioned markers or "*markings-out*," they remind us of the distinctive role that concepts have in every philosophical exploration. As stated in the introduction, the task in the foregoing sections has been to put such markings-out into play in a way that could safeguard their flexibility, indefiniteness and dynamic disposition. In addition to being the originator of this term, and despite his critical attitude to the aforementioned Marion, philosopher, Andrew Rawnsley reminds us about something important concerning the context in which markings-out is included:

> "Concepts" and "categories" are not the objects of philosophy, nor even of thought, they are the way in which we mark out a region to work with, work in, to work-out-of. The more flexible and fluid these markings-out become, the richer the region thus marked out.

> Philosophical activity is richest when it seeks to dissolve the rigidity
> of precise definition in favour of the opening up of regions of inves-
> tigation whilst remaining rooted in the life in which such activity
> always occurs and in which it has its sense.[51]

The expression "the richer the region thus marked out" reminds us of
a paradoxical possibility. Namely that in certain circumstances we can
refrain from conceptual rigidity and precision in order to experience life
and the world both with greater clarity and as richer in meaning. According to Rawnsley, we are also and simultaneously challenged to remain
"rooted in the life" in which this exploring activity takes place. In this
regard, I choose to share some experiences from encounters with suffering human beings. That is, persons who suffer from non-life-threatening
and so called medically unexplained conditions, who have *not* found
answers or adequate explanations in medical diagnostic menus and who
are beginning to turn in other directions and seeking ways to express
and understand how it is to be an afflicted, suffering human. The les-
son that may be learned from these encounters is that precisely designed
definitions of concepts, diagnostic clarity and listing and classification of
symptoms are assigned a subordinate role if the preconditions are present
for individuals themselves to participate in an exploration of a flexible
language for experiences of brokenness. These persons are likewise fully
capable of overruling the provisions of the diagnostic menus, provided
that they themselves are given space to express their confusion, aporia or
the undecidability of their life situation. One might expect and fear that
such processes leave the afflicted persons in a knowledge vacuum. However, what is at stake here is not a matter of objective facts about one's
life that satisfy the requirements for positive (scientific) certainty. Contact
with, and sharing of, some meaning-bearing and meaning-giving expressions of life that concern what it is to be a living person may instead invite
a *negative certainty*, i.e. a certainty about that for which one cannot possess positive certainty. As Marion indicates, such certainty is based on
our retranscribing and returning the phenomena that have hitherto been
objectified "into primordially given phenomena, because they are giving
themselves in themselves."[52]

Notes

1. Jean-Luc Marion, *Negative Certainties*, trans. Stephen E. Lewis (Chicago: The University of Chicago Press, 2015), 159 ff.
2. Oxford English Dictionary, "Object," in *Oxford English Dictionary* (Oxford University Press, 2018), https://en.oxforddictionaries.com/definition/object
3. Jean-Luc, *Negative Certainties*, 253.
4. See Thor Eirik Eriksen, Anna Luise Kirkengen and Arne Johan Vetlesen, "The Medically Unexplained Revisited," *Medicine Health Care and Philosophy* 16, no. 3: 587–600. (Dordrecht: Springer Science + Business Media B.V., 2013).

5. Jean-Luc, *Negative Certainties*, 156.
6. Ibid, 159.
7. Marion elaborates on this in the following question: "If in metaphysics we presuppose that entities may be divided into objects and nonobjects according to two absolutely heterogeneous modes of phenomenality, couldn't we instead try to suture this unjustified distinction by considering the opposite hypothesis: *that starting with a single and univocal phenomenality, the phenomena end up by diverging into objects and nonobjects according to the variations that they introduce into the dimensions of the same, unique phenomenality?*" Jean-Luc, *Negative Certainties*, 161.
8. Jean-Luc, *Negative Certainties*, 159.
9. This relates to Marion's hypothesis regarding a "univocal phenomenality that allows us to pass by gradations from the object to the nonobject, and back." Jean-Luc, *Negative Certainties*, 161.
10. My point, however, is not to determine exactly *where* such phenomena fall on a continuum, but more crucially to show that one and the same phenomenon can occur, as Gschwandtner describes it, "in manifold degrees and levels of saturation of givenness." Christina M. Gschwandtner, *Degrees of Givenness: On Saturation in Jean-Luc Mairon* (Bloomington, IN: Indiana University Press, 2014), 193.
11. "Online Etymology Dictionary." © 2001–2015 Douglas Harper, www.etymonline.com.
12. Gabriel Marcel, *The Philosophy of Existentialism* (New York: Citadel Press, 1984).
13. Medical analysis and investigation rely on counting and sorting symptoms, measuring health parameters, clarifying possible diagnoses and the launching of causal hypotheses.
14. Marcel, *The Philosophy of Existentialism*, 13.
15. Ibid., 14.
16. Gabriel Marcel, *The Mystery of Being: Volume I: Reflection and Mystery* (South Bend, IN: St. Augustine's Press, 2001), 103.
17. Katharine M. Larsson, "Understanding the Lived Experience of Patients Who Suffer from Medically Unexplained Physical Symptoms Using a Rogerian Perspective," PhD diss., Boston College, William F. Connell School of Nursing, USA, 2008.
18. That is, a translation that usually would assume a distinction between the following: (a) physical and mental pain; (b) acute and chronic pain; (c) local, regional and widespread pain; and (d) neuropathic (in the nervous system), nociceptic (tissue injury) and idiopathic pain (inexplicable).
19. Gabriel Marcel, *Being and Having*, trans. Katharina Farrer (Westminster: Dacre Press, 1949), 100.
20. Ibid., 101.
21. Ibid., 117.
22. Martin Heidegger, *Being and Time*, trans. Joan Stambaugh (Albany: State University of New York Press, 2010).
23. Ibid., 183.
24. Ibid.
25. Ibid., 267.
26. Katherine Withy, *Heidegger on Being Uncanny* (Cambridge: Harvard University Press, 2015), 8.
27. Ibid., 234.
28. Heidegger, *Being and Time*, 327.
29. Ibid., 182.
30. Withy, *Heidegger on Being Uncanny*, 61.

31. Ibid., 67.
32. Ibid., 53.
33. Martin Heidegger, *Zollikoner Seminars. Protocols-Conversations-Letters*, M. Boss, ed. (Evanston, IL: Northwestern University Press, 2001).
34. Heidegger, *Zollikoner Seminars*, 132.
35. Jean-Luc Marion, *Being Given. Toward a Phenomenology of Givenness*, trans. Jeffrey L. Kosky (Stanford, CA: Stanford University Press, 2002), 74.
36. Ibid., 162.
37. Marion's philosophical theory of the flesh is primarily inspired by the French philosopher Michel Henry.
38. Jean-Luc, *Being Given*, 125.
39. Ibid.
40. Ibid., 231.
41. Ibid.
42. Ibid.
43. Ibid.
44. Henry, Marion's main source of inspiration concerning the body, says the following in relation to this: "Through affectivity, absolute Life comes into oneself, is experienced and enjoyed. Affectivity is not a fact or a state, it is not one property among others." Michel Henry, "Incarnation and the Problem of Touch," in *Carnal Hermeneutics*, ed. Richard Kearney and Brian Treanor (New York: Fordham University Press, 2015), 134.
45. Jean-Luc, *Being Given*, 317.
46. Ibid., 316.
47. An example would be the concept symptom. See Thor Eirik Eriksen and Mette Bech-Risør, "What Is Called Symptom?" *Medicine Health Care and Philosophy* 17, no. 1 (February): 89–102 (Dordrecht: Springer Science + Business Media B.V., 2014), doi: 10.1007/s11019-013-9501-5
48. *Dictionary of the Danish Language, 1750–1900* (Danish: Ordbog over Det Danske Sprog, 1750–1900) (Cophenhagen: Det Danske Sprog-og Litteraturselskab (DSL)). http://ordnet.dk/ods (accessed February 1, 2018) (my translation).
49. Heidegger, *Being and Time*, 108.
50. Daniel O. Dahlstrom, "Heidegger's Method: Philosophical Concepts as Formal Indications," *The Review of Metaphysics* 47, no. 4 (1994): 775–95, 780.
51. Andrew Rawnsley, *Practice and Givenness: The Problem of "Reduction" in the Work of Jean-Luc Marion* (Oxford: The Dominican Council and Blackwell Publishing, 2007), 708.
52. Jean-Luc, *Negative Certainties*, 202.

7 No Way Out
A Phenomenology of Pain

Christian Grüny

Introduction

Pain is a complex phenomenon. What does a sprained ankle have in common with a testicular torsion or a bullet wound, and how do these relate to labor pains, chronic neuralgia or the suffering of a torture victim? Does it make sense to use the same term for these radically disparate experiences? Still, we don't hesitate to call all of them "pain" and even extend the use of the word to emotional suffering. Speaking of the pain of loss is not a metaphor, and separation and death *hurt* just as much as physical injury—albeit in a different way. We don't usually confuse social injuries with physical ones, but the facts that no physical pain is emotionally neutral and that in somatization emotional anguish can be expressed by bodily symptoms make it clear that the line between the physical, the mental and the social cannot be clearly drawn. So rather than saying that this chapter will be dealing exclusively with physical pain, we should say that it focuses on pain of physical *origin*.

In the following, I will attempt to flesh out a phenomenology of pain by drawing on Merleau-Ponty's phenomenology of the body, Erwin Straus's concept of sensing, and Bernhard Waldenfels's philosophy that retains the idea of the lived body but focuses on rupture and non-coincidence within experience. After considering the problem of the unitary character of the phenomenon of pain, I will turn to a concept of pain as process and the question of the experience of the body in pain. In conclusion, I will take a brief look at questions of meaning.

At Pains

When we look at different experiences, it is interesting to see what distinguishes the examples I gave in the first section. First of all, the intensity of pain varies greatly from the hardly noticeable to the unbearable. Secondly, pain takes on different forms, and we speak of a stinging, a gnawing, a burning pain etc. It's a third difference, however, that is most important: a difference in meaning and context. Slightly bruising your shoulder or breaking it causes pain of different intensities, but they also impinge on your life in very different ways, and if after surgery and a prolonged healing process a chronic pain condition remains, the experience and its impact change completely. While the extreme pain of giving birth may be somewhat alleviated by the fact that it is transitory and gives life, so to speak, the pain of torture is aggravated because it is intentionally inflicted and there is no telling when it will end. So while it is extremely important to include these meanings and contexts into a philosophical inquiry into the nature of pain, it might still make sense to look for an experiential core that warrants the use of a single term without losing sight of the vast differences it is meant to encompass.

But pain is complex in another sense that threatens the very idea of such a core: far from being a clearly defined single phenomenon or the simple transmission of information concerning a physical lesion it was once thought to be, pain involves the whole nervous system or rather the whole organism. To quote Nikola Grahek: "although pain appears to be a simple, homogenous experience, it is actually a complex experience comprising sensory-discriminative, emotional-cognitive, and behavioral components."[1] Note that Grahek does not refer to neurological or physiological facts but to components of experience (which are, to be sure, related to such facts). While it seems undisputable that pain comprises these different *dimensions*, it might be questionable to construe them as *components*. Is it appropriate to say that experience consists of parts? The logic behind such an understanding is usually founded in some reference to biological structures and functional systems, but Grahek's argument is a different one: there are pathological conditions like pain asymbolia where the sensation of pain is dissociated from the emotional and behavioral aspects, *ergo* they have to be independent.

But is this plausible? His point is to defend the complexity of the experience against a view that identifies pain with a certain quality or quale of sensation. But the best way to counter such a reductionist stance seems to me to point out that this quality is an *abstraction* from a complex experience rather than a component of it—which is precisely what the term *complexity* entails. Normally this quality simply doesn't exist by itself, and when it does, as in pain asymbolia, it constitutes a radically changed experience that has little to do with pain as we know it or, for that matter, with any other experience.

Insisting on complexity is "radical[ly] antisubjectivist" only if subjectivism consists in postulating an irreducible and defining quality that can be studied without any reference to its context. To claim that "the sensation of pain or pain quality plays no important role in our total pain experience and that what really matters is only how we respond affectively, what we believe, and how we act,"[2] throws the baby out with the bath water. If there is something that can rightly be called "pain quality" it is an experience in an affective and social context, an experience whose quality is obscured by drawing a sharp distinction between sensation and response. To insist on the complexity of pain, as Grahek does, should amount to a defense of experience against subjectivist reductionism.

On the other hand, an objectivist position that hopes to reduce pain to scientifically observable facts is just as misplaced. Since, as Geniusias observes, "the pre-scientific experience of pain is the very subject matter of diverse sciences of pain and . . . scientific determinations are meant to be nothing other than clarifications of pain experience,"[3] this pre-scientific experience must remain the guiding thread of any scientific research. As a complement to this research we need an approach that tackles experience head-on without ignoring the results of scientific research or falling into the trap of subjectivist reductionism.

The key to such an approach is an understanding of pain not as a thing, a state, or a conjunction of states but as a process that involves the whole person and whose complexity lies in the way it implicates all kinds of different biological structures and layers of meaning so that it cannot be easily mapped onto distinctions like that between sensation and feeling. Even hybrid categories like Carl Stumpf's "feeling-sensation" (Gefühlsempfindung)[4] are not enough if, as two of the most important medical researchers on the subject remark, "pain becomes a function of the whole individual, including his present thoughts and fears as well as his hopes for the future."[5] There really is an irreducible experiential quality here but it lies in the complex process as a whole, and accordingly the quality changes if any of its dimensions change.

In my view, the philosophical approach best suited to this task is still the phenomenology of the lived body as it was elaborated by Maurice Merleau-Ponty in his *Phenomenology of Perception* and related writings. Insisting on the lived body as a "third term between the psychic and the physiological"[6] while drawing on both physiological and psychological research, Merleau-Ponty offers a concept of experience and the body that is decidedly anti-reductionist. Surprisingly, pain is mentioned in the *Phenomenology of Perception* only in passing and without any systematic consequences, so instead of simply presenting a phenomenological theory of pain we have to develop such a theory with the means that Merleau-Ponty provides us with.

In addition to this, there are two other thinkers I will mainly rely on: Erwin Straus and Bernhard Waldenfels. In his *Vom Sinn der Sinne* (*On the Sense of the Senses*), Straus presents his concept of sensing (*Empfinden*) as an elementary mode of interacting with the world, which comes to play an important role in the *Phenomenology of Perception*. Straus's original elaboration provides a necessary supplement for an adequate account of the complex experience of pain. Bernhard Waldenfels, on the other hand, has developed the phenomenology of the body in important ways. Waldenfels is most known for his phenomenology of the alien (*Phänomenologie des Fremden*), and only few of his books have been translated into English. His philosophy is centered on the notions of non-coincidence, diastasis and asymmetry, which are highly relevant for an understanding of the body in pain, which will in turn shape our understanding of the lived body in general.[7]

Motor Physiognomy

Merleau-Ponty begins the second part of his *Phenomenology of Perception* with a long chapter on *le sentir*, which Colin Smith renders as "sense experience." This translation misses one of the author's essential points and also obscures his reference to Straus, Max Scheler, and Heinz Werner. Like these, Merleau-Ponty chose the verbal form to point to the fact that the basic embodied sensory experience must be conceived as a dynamic process: sensing instead of sensation, *Empfinden* instead of *Empfindung*. According to Straus (and Merleau-Ponty), the hypothetical qualia of sensations are products of their assimilation to physiology on the one hand and to knowledge and cognition on the other, thereby producing the idea of an entity that is at once a physical occurrence and a unit of information. Getting rid of this artifact is a prerequisite for formulating an adequate theory of experience.[8]

Straus describes sensing as an embodied, affectively charged mode of interaction or communication with the world that must be distinguished from cognition in that it lacks any reflective detachment. Rather than detached observation, it is sympathetic experience (*Erleben*) that cannot be separated from the movement of the body. Whatever I encounter affects or threatens or promises to affect me in a certain way, and it prompts a certain type of movement toward or away and a specific posture. The different senses and also pain must be considered variations of "the basic theme I-and-the-world,"[9] ranging from harmonious communion to threatening violence. This primordial mode of interacting with the world isn't a phase in ontogenetic or phylogenetic development but a substructure of all experience that comes to the fore in certain situations. Straus draws a clear distinction between sensing and perception because he assimilates the latter to an objective cognition that encounters "a world of things with fixed and variable properties in a general, objective space and a general, objective time."[10]

This is where Merleau-Ponty's conception diverges from his: for the philosopher, this kind of objectivity belongs to scientific theory, and projecting it onto perception is a scientist myth. Despite the increase of differentiation and integration from the primordial mode of sensing to fully developed perception, there is a fundamental continuity between the two, hence: "Every perception takes place in an atmosphere of generality and is presented to us anonymously."[11] There is no clear break from sensing to cognition, just as there is no break between the lived body as "natural self"[12] and the self-conscious ego.

If our primary mode of encountering the world is sensing, conceived not as a reception of data from the world but as an embodied interaction with it that encompasses movement and affectivity, then pain must primarily be an occurrence within this sphere; only here can it make sense to speak of "a perception inclusive of sensation, emotion and cognition."[13] But if any sensory quality is a specific form of this interaction, what is the form of interaction in pain? Merleau-Ponty's main example for a specific instance of sensing is color, even though this dimension of experience is hardly noticeable in everyday perception where colors appear as properties of defined objects or self-enclosed qualities. He observes: "Sensations, 'sensible qualities' are then far from being reducible to a certain indescribable state or *quale*; they present themselves with a motor physiognomy, and are enveloped in a living significance."[14] It is this idea of motor physiognomy that I find particularly well suited for conceptualizing pain. Merleau-Ponty draws on research from early twentieth-century Gestalt psychology that suggests that colors embody a certain movement impulse and are only fully realized in perception when traces of this movement occur. If this is true, even the standard case of qualia in cognitive science and the philosophy of mind must be understood as an occurrence between a sensing body and an environment where quality and affect cannot be separated.

While this might seem speculative in the case of colors, it is highly plausible in the case of pain. The only problem with Merleau-Ponty's descriptions of the relation to the world in sensing is that they are remarkably harmonistic, and the metaphors he employs are communication, pairing, synchronizing, communion—it seems like nothing bad ever happens to the lived body in this colorful, friendly world. We can retain from these metaphors the idea that in any instance of sensing a quality appears *in* the interaction, involving both the world and the embodied subject, but we have to question the harmonious to and fro they suggest. All sensing includes "an experience of *being moved*,"[15] or, more precisely, an experience where moving and being moved cannot be clearly separated, and in pain this balance is shifted drastically toward the passive: something is being done to us, even if it is done by our own hands. But we have to turn to other thinkers if we want an adequate account of this shift.

There is an observation we find in numerous texts on pain that seems to disrupt the idea of a continuous movement between subject and environment altogether and that undermines the notion that pain is *primarily* an information about a certain state of things. In his book on pain, F.J.J. Buytendijk employs biological terms when he calls sudden pain "an unexpected rupture of communication between organism and milieu."[16] His reference to expectations shows that he is not talking about a fact stated by an outside observer but something that is felt by the organism itself. If communication stands for a more or less peaceful exchange, pain must indeed be felt as its rupture. But maybe we should speak more neutrally of interaction, which would avoid some of the normative overtones the notion of communication tends to have, and of pain as a rupture *in* rather than *of* interaction, a rupture that is itself a mode of interaction but radically transforms it. Sudden pain shares this trait with other sudden occurrences like loud noises or flashing lights in that it momentarily eclipses all other perceptions—it catches us off guard. Pain differs from these because its demand to attend to it does not cease, and it seems to me that all pain has this element of rupture, of a disruption of the unproblematic flow of experience that we can never be adequately prepared for. Even minor injuries that briefly capture our attention but can then be ignored continue to gnaw at the edge of consciousness.

The concepts Waldenfels suggests to subvert the traditional phenomenological notion of intentionality from within are particularly elucidating here. Intentionally referring to something presupposes having been affected, for which he suggests the Greek *pathos*. Any act we perform must then be considered a response to this pathos, an affection that never appears in itself but only in the response that regards it *as something*. The pair of pathos and response is thus not identical with that of stimulus and reaction: "Pathos and response do not follow one after the other like two events; they are not even two distinct events, but one and the same experience, shifted in relation to itself: a genuine time lag."[17] Waldenfels calls this time lag *diastasis*, which does not refer to a delayed reaction or a measurable temporal distance between cause and effect but to the fact that we're responding to something that has always already happened, that we're always too late in our intentions and our making sense of something.

This dimension of pathos and diastasis tends to get normalized and finally all but disappears in habitual everyday existence; only in experiences like surprise, shock, pain, and trauma does it appear as disruption and disturbance. Pain continues to disturb. Again and again it lets our attention slip from whatever we're doing and draws it toward itself. We cannot get over it because something continues to affect us and to demand our response, something we never quite come to terms with. It does that because something continues to happen that we cannot escape and, what's more, inevitably have a part in. This is what we mean when we say "it hurts": an infliction and intrusion that continues to act upon us and forces us into re-acting by unavailingly attempting to withdraw.

In the case of injury by an outside factor or agent like the blade of a knife or a hot stove this recoiling impulse takes the shape of an actual physical movement, but in the case of internal pains no such movement is possible. In one of his few remarks on the subject, Freud claims that pain acts "like a continual instinctual stimulus, against which muscular action, which is as a rule effective because it withdraws the place that is being stimulated from the stimulus, is powerless."[18] I would argue that this powerless attempt to get away is *part* of the pain rather than merely a reaction to it, and it is precisely the fact that it is impossible that makes it so upsetting. This, the futile effort to withdraw from part of ourselves, is the motor physiognomy of pain, and it obviously encompasses the sensory, affective, and behavioral aspects Grahek insists on. Freud underplays how catastrophic this can be when he continues: "If the pain does not proceed from a part of the skin but from an internal organ the situation is still the same. All that has happened is that a portion of the inner periphery has taken the place of the outer periphery."[19] To treat an inner organ as one's periphery amounts to an internal fragmentation, a split that runs right through the embodied self. To really flee from the intrusion of pain we would have to flee from ourselves.

But still: wherever the pain is situated and whether it has a perceptible cause or not, it is not a worldless event. Sensing is our primordial mode of encountering and dealing with the world, neither a self-enclosed quality nor an emotion but a type of interaction. We respond to being affected in a certain way, and this response contributes to how the world we encounter is perceived. So when, as Straus writes, "in pain the world encroaches upon us and subdues us,"[20] it is of secondary importance whether there is an actual external cause. We feel that we are being assaulted and find ourselves exposed. The greater its intensity, the less we can ignore this assault and our involuntary response to it, and we find ourselves continually being thrown back from the world of perception with its things, qualities and actions to the less differentiated sphere of sensing where we are at the mercy of a hostile world that may even have occupied our insides. This feeling is increased immeasurably if the pain is actually inflicted on purpose by somebody else. In this respect, pain isn't so much a means of torture but its essence.[21]

In its demand for attention pain has a totalizing tendency, and in extreme cases it can occupy all experience, reducing the world to nothing but an undifferentiated origin of assault. Since withdrawal is impossible but still attempted desperately, organized and purposeful behavior tends to disintegrate into what Kurt Goldstein called "catastrophic reaction": "disordered, inconstant, inconsistent, and embedded in physical and mental shock."[22] To be sure, this totalized disintegration only happens in extreme cases. Agustín Serrano de Haro distinguishes "invasive," "co-attended" and "inattended" pain and stresses that even in situations of extreme pain some background awareness of the situation and one's body remains.[23] Disintegration is a tendency, not a given. But if the pain persists, the world of the sufferer is changed permanently even if he/she retains the outside appearance of self-control. A typical description of this situation is that "terrible things are being done to the person and worse are threatened" and "others, or outside forces, are in control and the will is helpless."[24]

If, as I have claimed, pain is not just something that happens to us but something that we *do*, however reluctantly, we must be able to change it by changing our behavior. Indeed, intensifying the futile effort to withdraw by enacting it with our whole body makes the pain worse, while many of the non-invasive methods of treating it include relaxation techniques. Since there is no *adequate* bodily reaction to the assault of pain, we must try to counteract our own tendency by practicing what Buytendijk calls "the somatic equivalent of composure."[25] Our emotional situation and our knowledge and beliefs also have a distinctive impact. Being emotionally unstable or outright depressed and expecting the worst considerably intensifies pain, and Geniusias rightly observes that "bodily, emotive and cognitive responses up to a large degree make up the painfulness of pain."[26] Unfortunately the inherent tendency of pain is to provoke the worst responses possible, and here too we must counteract our own impulses to fight it and rely on the help of others to provide care, emotional support and information. In the case of minor pains our response remains local and brief, and we continue with our lives without further ado, but even here we find the same basic structure. To uncover it, we have to look at pains of greater intensity.

Most pains take their place in our ongoing intentional relation to the world, and we all have developed techniques to hold their demand for attention at bay (and of course there are always painkillers within reach). But what about the intentionality of pain itself? Does it make sense to call pain intentional? Merleau-Ponty described sensing as a primordial mode of intentionality that cannot be grasped with the Husserlian triad of *ego*, *noesis* and *noema*. Intentionality then refers to an elementary type of dynamic relatedness where any quality "does not rest in itself as does a thing, but . . . is directed and has significance beyond itself."[27] In this type of intentionality the quality does not reside in an intended object but in the relation itself, as I have tried to elaborate. Mistaking pain for an intentional or mental *object* amounts to assimilating the sentence "I feel pain" to "I feel a sharp knife." Guy Douglas illustrates this fallacy with analogous examples: if "to leave in a hurry" was structurally the same as "to leave in a taxi,"[28] a hurry would have to be a means of transportation. Instead, it must be understood adverbially: leaving in a hurry is a *mode* of leaving, just like feeling pain is a mode and not an object of feeling. As such, it may tinge all our conscious relations to the world, as in Sartre's observation that "pain can itself be indicated by objects of the world."[29] The global character of sensing is thus transformed into an atmosphere that colors everything.

But of course that is not the end of the story. When we consciously assume a certain relation to our pain, we are not *in* it anymore, as it were, but actually do relate to it as an intentional object, and this is far from exceptional: we do it all the time. Abraham Olivier makes the important point that if pain was a state or mode and nothing else this would mean that it was truly inexpressible. We would then "either talk about something else but not pain when we talk about pain, or . . . revert to a position prior to language in which there is no talk and only pain."[30] In fact relating to pain this way often has a liberating quality: it makes a global experience into a localized event that we can distance ourselves from and thus try to master. We can describe how it feels and specify the kind of help we need.

We find a famous example of this kind of objectification in Nietzsche's *Gay Science*: "I have given a name to my pain and call it 'dog': it is just as faithful, just as obtrusive and shameless, just as entertaining, just as clever as any other dog—and I can scold it and vent my bad moods on it, as others do with their dogs, servants, and wives."[31] Naming the pain creates distance, and naming it "dog" aims to make it into a subservient being. Nietzsche's cynical account obscures one thing, however: distance and mastery remain precarious. The stronger the pain gets, the less likely our attempts to objectify it will succeed. So the point is not that this is an impossible or inauthentic way of relating to pain. What's important is to understand that it is not our primary way of feeling pain. We never get a clearer understanding of the experience of pain if we insist on conceptualizing it as an intentional or mental object because this object is really a process we are implicated in and can never fully detach ourselves from.

Not a Single Body

If pain is a specific instance of sensing, a futile attempt of withdrawal from a hostile intrusion, how does it affect our experience of the body? What is a body in pain? *The Body in Pain* is of course the title of a much-quoted book by Elaine Scarry that attempts to construct a fairly speculative theory of culture from an analysis of the experience of pain, focusing on the extreme, socially destructive pain of torture.[32] While her perspective is illuminative in many ways (and problematic in others), she doesn't have a lot to say about the actual experience of the body in situations of pain, apart from its "huge, heavy presence"[33] that increases as that of the world diminishes. In order to develop a more nuanced account of the body in pain, we have to take a step back and ask how the body is experienced in everyday experience and then examine how this changes in pain. For this I will turn back to Merleau-Ponty.

As is well known, the *Phenomenology of Perception* is first and foremost a philosophy of the body—but there is a reason why it has its title. The phenomenal or lived body, the *Leib* Merleau-Ponty is concerned with, is a body in action and in perception, and the most important concepts that characterize it are situatedness and involvement. The body marks a perspective on the world, a place from which we act and perceive that cannot be reduced to an objective location in space; instead, phenomenal space is constituted by a multitude of relations that are not static perspectives but related to modes of action, of bodily involvement in the world. The lived body cannot be understood apart from this involvement but is determined by it through and through.

The psychological concept that best captures this is the body schema. As Merleau-Ponty traces it, the idea of the body schema was developed from a kind of mental representation of the relative positions of the limbs to the point where "the body schema is finally a way of stating that my body is in-the-world."[34] Thus identifying his own concept of the lived body with a sophisticated version of the body schema, he makes it clear that he understands the lived body not as an entity, not as a complex of organs and parts but as a system of different ways of relating to the world. The systematic character is important to this argument, and Merleau-Ponty carefully stresses its unity, which he compares to that of the work of art. The body in action doesn't have to be assembled from parts; its division into discrete parts is a posterior move, the division of a system where every "part" is implicated in the others and they are all unified in their orientation toward worldly goals. It would be misleading, however, to understand this coherence as that of a whole, as an integral unity where everything has its place or rather: where there is an "everything" that it could encompass.

Intentional through and through, the lived body in action is discreet rather than discrete, and it is so to the point of being self-effacing in the literal sense of the word. This understanding of the lived body can be called adverbial, designating a mode of being, a functional structure rather than a substantial entity, which makes it somewhat misleading to speak of "the body" at all. The *Phenomenology of Perception* abounds with phrases like "bodily existence," "bodily experience" or "bodily space," and this adjective captures Merleau-Ponty's intentions much better. In this adverbial or modal existence, the body itself tends to disappear: it is, as Drew Leder appropriately titled his book, an *absent body*.[35] The body or parts of it may be absent from experience because the focus of our attention is directed somewhere else, but they may also disappear because they are at the very center of attention or, more precisely, because they are the means of our attentiveness: when I write it's not just my feet, the back of my head etc. that disappear but my writing hand as well. Leder speaks of "background disappearance" and "focal disappearance" and adds a third type that characterizes our non-awareness of our inner organs: "depth disappearance."[36] To be sure, there is always a background awareness of our body, but it is stunning how little we notice even those parts of it that are within our visual field. We could say that the ideal kind of presence of our body is that of a neutral "I can"[37] where the body is a reliable and flexible system of abilities that never gets in the way, rather than a perceptible whole. Ultimately one would have to say: "I am a field, an experience."[38]

Recognizing and describing this adverbial or modal nature of our body is one of the most important achievements of Merleau-Ponty's philosophy, but it does tend to obscure other modes of relating to it.[39] First of all the body has to be trained and groomed, washed and clothed, in short:

explicitly addressed and acted upon, in order to function at all. In some of these activities it is treated almost but never quite like a thing. The most fundamental and quotidian way of relating to one's own body is touching it, and the most puzzling instance of this is touching one hand with the other, which Merleau-Ponty calls, following Husserl, "a kind of reflection."[40] The touching and the touched become interchangeable but never quite coincide, and "in this bundle of bones and muscles which my right hand presents to my left, I can anticipate for an instant the integument or incarnation of that other right hand, alive and mobile, which I thrust toward things in order to explore them."[41] In his later philosophy, this relation is formalized and universalized: the touching of one's own hands becomes the paradigm of the chiasm and the key to our relation to the world where the body is the hinge between the touching and the touched, seeing and seen, self and other, and self and world.[42]

But what happened to the "bundle of bones and muscles"? What about the materiality that can't be incorporated or formalized? What about the body being exposed to hurt and injury? The body that tends to disappear in perception reappears when things go wrong, when we sprain our ankle, when a sudden headache hits us or we are struck by a falling branch. This mode of appearance of the body, which Leder calls "dys-appearance,"[43] has some obvious parallels to the appearance of the broken useful things in Heidegger's analysis: the focal disappearance of my writing hand will only last until it starts to hurt, just like the pencil will only remain handy until its tip breaks.[44]

Pain seems to be a paradigmatic case of this kind of dys-appearance, which Merleau-Ponty largely ignores. It is obvious that by drawing attention to itself, pain also lets the affected body part appear. But in what way, or *as what*? First of all, the affected organ is ejected from the system of implicit functioning and obtrudes on me. It changes from a transparent medium of action or implicit background into an impediment. Its spatiality changes: whereas the hand as a medium of exploration is in motion as a whole, a vector rather than a definite location even when it is momentarily resting on the table, pain is localized and has a localizing effect. The cramped muscle is clearly located, and kneading it treats it as a material object. The kidney that suddenly makes itself felt emerges from darkness, letting something appear where there used to be nothing but "heavy mass."[45] Apparently this is even true for the heart, and according to the physician and philosopher Herbert Plügge the first words of the cardiac patient are these: "I never knew that I have a heart. Now I know."[46] Even though the heart is one of the few inner organs that can be felt all the time, pain seems to change this experience so drastically that it appears to produce something that wasn't there before at all, a thing inside oneself that remains a foreign body, as it were.

Numerous authors have associated this experience with materiality, insisting that it reveals the material dimension of the body. I think this is an accurate description, even though it remains unclear what exactly could be meant by materiality. Plügge speaks of a kind of "coagulation" or "congealing" of the lived body and then writes: "The res extensa appears as phenomenon at the heart of the normally unnoticed lived body."[47] There is obviously something fundamentally wrong here: first of all, Descartes' concept of *res extensa* explicitly made no reference to materiality, density etc. but referred to pure spatial extension. Second, it is a theoretical construct that by definition can't be experienced as such and that philosophers have been keeping alive long past its usefulness in science.

A more productive phenomenological conception of materiality can be found in the second volume of Husserl's *Ideas* and in some of Hans Jonas' essays. Husserl points to the fact that we discover the materiality of the world not by watching it but by feeling its resistance. "Impact and pressure" let us experience the firmness and the weight of the things around us, and we have to strain our muscles to explore them.[48] The "I can" is intimately related to an "I cannot" where the resistance of the world is too great for our powers or where we feel its potentially crushing impact[49] and while Husserl primarily speaks about the materiality of things, it is obvious that our own body shares this property. However, we should not misunderstand the feeling of resistance to always imply suffering and failure: there can be joy in exerting our physical powers and overcoming obstacles, and even the enjoyment of the fluidity of our bodily movements at the height of its capability could not exist without a trace of the feeling of weight and resistance.

Still, there is a degree of heterogeneity here that comes to the fore in pain and that cannot be explained away. Pain is the paradigmatic experience of a force that we suffer and that forces us into awareness of our own materiality. It limits our range and lets our body get in our way. If we are a thing among things, "bundles of bones and muscles," we can get crushed between them. If my reconstruction of the motor physiognomy of pain is feasible, interior pains without any perceptible external cause give us the same experience of force and vulnerability. This might be one of the reasons why we tend to employ the "language of agency,"[50] as Scarry calls it, to describe all kinds of pain: my abdominal pain feels as if I was being stabbed not just because this is my way of externalizing and objectifying a private experience, as she thinks, but because I really do feel assaulted by an alien force that alienates part of my body, and I find myself nailed down without any way out.

The body is obviously both, disappearing capability and vulnerable materiality, but how can we reconcile these two dimensions? Can we simply add materiality to the concept of the lived body? Is there an overarching concept that encompasses both? The experience of pain and hurt throws this into doubt. "The body in dys-appearance is marked by being away, apart, asunder,"[51] Leder writes, and the very fact of dys-appearance is a reminder that the body is not one even when it is not broken apart.

We can turn to Waldenfels one more time for an elucidation of this inner heterogeneity. There is a sentence from Merleau-Ponty's *The Visible and the Invisible* that sums up his own approach so well that he quotes it several times: "the originating (*l'originaire*) breaks up, and philosophy must accompany this break-up, this non-coincidence, this differentiation."[52] This means that there is no originary unity of the body just like there is no originary unity of being. Waldenfels spells out the formal chiasm of the touching/touched into the concept of a body that is destined to miss itself as it relates to itself, whose very reference to itself (*Selbstbezug*) is also an evasion of itself (*Selbstentzug*): "This noncoincidence should be viewed as a liability, for it characterizes the very being of our body, which refers to itself and at the same time evades itself."[53]

While Helmuth Plessner's distinction between being a body (*Leib*) and having a body (*Körper*)[54] is one of Waldenfels' points of reference, this might create the impression that there are exactly two modes of existence of the body that we somehow have to mediate between. But having a body refers to an objectifying relation to our body that ranges from cutting one's fingernails to neurophysiological research, and while it may make formal sense to group all these different ways of relating to ourselves together, this should not obscure their heterogeneity: taking care of our body is not the same as reifying it. The materializing effects of pain disrupt or transform the implicit lived body in a different way, and they find no place in Plessner's distinction at all.

Moreover, all these transformations usually remain local and transient, they are never complete and stable enough to form "a body." Not even scientific objectification creates another version of the body that we "have": while it obviously leaves permanent traces in our self-understanding—I *know* that I have a heart, roughly how it works and what it looks like—it will never completely take root within the lived body, let alone replace it. If it did, the strange statements of the cardiac patients would be utterly inexplicable. Pain is the experience that makes the non-coincidence at the heart of ourselves most acutely felt.

Centering our understanding of the body not on a substance or unitary organization but on non-coincidence prevents us from declaring one of those modes the original core and others as negligible or improper—but it doesn't absolve us from carefully investigating the different modes and their relations. The body is a field of different modes of experience that are related but not identical. The implicit, adverbial lived body that merges into its relation to the world can be seen, touched, hurt, dissected, and reconstructed from a third-person perspective; it is one body, but this body is not one.

The What, the Why and the How

There is, of course, a lot more to be said about the experience of pain, for instance about the way it affects space and time and our self-perception, and how all this changes again in the cases of chronic pain and torture. Also, the concept of a pluralized body had to remain a mere sketch. To conclude, I would like to add a few equally sketchy remarks on the problem of meaning and sense—another field that would demand much more attention.

In its dependence on context, pain is always intimately related to questions of meaning, and the plurality of contexts makes it clear from the outset that asking for "the" meaning of pain makes little sense. In the case of excessive pain deliberately inflicted by someone else, there is no meaning to be found outside the pain because it is its very embodiment. Torture primarily aims at destruction, and destruction is the meaning of the pain: its potentially destructive effects are intentionally mobilized and the victim "understands" this meaning only too well.[55]

In chronic pain, the relation to meaning is more complex. While normally the first and simplest questions about pain's meaning and sense— why does it hurt? When will it stop? What can be done about it?—are fairly easy to answer or at least presumed answerable, in chronic pain they all become problematic. The mere fact of not knowing when the pain will stop will exacerbate it, and not having satisfactory means to fight it makes things worse. Almost inevitably the question will be transformed and shifted to other kinds of meaning: psychological, moral and even metaphysical.

Finding metaphysical meaning presupposes faith in some kind of order in the world, an order where everything has its place and there is a reason for everything that happens, whether there is a supreme will behind it or not. This kind of faith can be surprisingly effective but it isn't as widespread as it used to be—and it cannot be prescribed. The moral dimension that often goes along with this but can also be linked to a more secular understanding may help bear the burden of chronic pain, but it is far from harmless: there must be something I have done to deserve this, and the pain is a punishment for my misdeeds even if I have no idea what they are or what to do about them.

Luckily prayer isn't medicine's last resort, and if the pain can't be assuaged pharmaceutically or surgically, other ways of treatment are employed, among them the relaxation techniques I referred to in the second section. But meaning tends to play a role as well, and biomedical treatment is complemented by hermeneutics. Psychosomatic approaches try to provide patients with the resources to understand the pain and its consequences on them, relate it to other issues in their lives, and transform its meaning to make it less damaging. This should not be confused with the attempt to give it meaning, as if a higher meaning would somehow elevate pain to a "meaningful" or beneficial experience that somehow promotes personal growth. When Geniusias insists on complementing the de-personalizing effect of chronic pain with a re-personalization, he means that we have to take account of the way the pain induces a change in personality that cannot be reduced to mere destruction;[56] when Olivier speaks of "liberation," he is not talking about a liberation *through* pain but a liberation *from* pain: what's at stake is "encountering hurt," "surviving affliction" and "overcoming agony,"[57] the difficult and precarious ways of coping with it.

But there's a thin line between restoring agency and blaming the victim. If the pain *per*sists despite all efforts to treat it, it must seem like the patient is *in*sisting on it, unwilling to let it go—if she isn't making it up. Indeed, this suspicion seems to be almost inevitable, as Arthur Kleinman remarks: "If there is a single experience shared by virtually all chronic pain patients it is that at some point those around them—chiefly practitioners, but also at times family members—come to question the authenticity of the patient's experience of pain."[58] It would be too simple to blame this on an exaggerated faith in the powers of modern medicine, even though this might be a contributing factor as well. It is likely that physicians will find it problematic to deal with their own powerlessness, but there is another factor that runs deeper: for most of us, pain is an episodic occurrence. It will pass. If somebody tenaciously claims that it doesn't, we have to strain our imagination to relate to it, which apparently isn't so easy. In chronic pain, Scarry's statement "to hear that another person has pain is to have doubt"[59] starts to ring true. What's at stake for the patient is the recognition of his/her own experience as valid and him/her belonging to a common social world.

This is a real problem that physicians should be aware of, but it is not the truth about pain as such. In most cases doubt is not a relevant category at all. Usually there is a direct link between the expression of pain and its perception by others: hearing someone moan or cry in pain and seeing them flinch affects us physically. We don't "hear that someone has pain," as if this was a neutral information that will yet have to be proved right or wrong; we hear and see someone in pain. We might take this as an example of what Merleau-Ponty called "intercorporeity" and linked to the chiasm,[60] but maybe this is still too formal. Emmanuel Levinas offers a different account: he identifies the body with exposition and vulnerability as he speaks of "the living human corporeality, as a possibility of pain, a sensibility which of itself is the susceptibility to being hurt, a self uncovered, exposed and suffering in its skin,"[61] thus placing at the center precisely that which Merleau-Ponty ignored. However, the fact that we are exposed and vulnerable doesn't imprison us within ourselves but opens us to the vulnerability of the Other. Our non-indifference toward him/her is dependent on our own bodily materiality.

Of course non-indifference is neither compulsion nor obligation; all it says is that turning away is not a neutral act. Pain's deepest meaning is always negative: that it should stop. It is a cry for help even if no cry is heard, and there are no innocent bystanders. It implicates the vulnerability of our material bodies in a way that there can be no neutrality. What it is asking for is ultimately not explanation or some sort of higher meaning but recognition and relief.

Notes

1. Nikola Grahek, *Feeling Pain and Being in Pain* (Cambridge, MA and London: The MIT Press, 2001), 2.
2. Ibid., 95.
3. Saulius Geniusias, "Phenomenology of Chronic Pain: De-personalization and Re-personalization," in *Meanings of Pain*, ed. Simon van Rysewyk (Cham: Springer Verlag, 2016), 148.
4. Carl Stumpf, *Gefühl und Gefühlsempfindung* (Leipzig: Barth, 1928).
5. Ronald Melzack and Patrick D. Wall, *The Challenge of Pain* (London: Penguin Books, 2nd ed. 1988), 145.
6. Maurice Merleau-Ponty, *Phenomenology of Perception*, trans. Colin Smith (London New York: Routledge, 2002), 140.
7. For an elaboration of these ideas cf. Christian Grüny, *Zerstörte Erfahrung: Eine Phänomenologie des Schmerzes* (Würzburg: Königshausen & Neumann, 2004).
8. Cf. Renaud Barbaras, "Affectivity and Movement: The Sense of Sensing in Erwin Straus," *Phenomenology and the Cognitive Sciences* 3 (2004). For a similar critique of the concept of sensation cf. Jean-Paul Sartre, *Being and Nothingness: An Essay on Phenomenological Ontology*, trans. Hazel Barnes (London: Routledge, 1956), 315.
9. Erwin Straus, *Vom Sinn der Sinne: Ein Beitrag zur Grundlegung der Psychologie* (Berlin: Springer, 2nd ed. 1956), 402.
10. Ibid., 215.
11. Merleau-Ponty, *Phenomenology of Perception*, 279.
12. Ibid., 239.
13. Abraham Olivier, *Being in Pain* (Frankfurt am Main: Peter Lang, 2007), 51.
14. Ibid., 243.
15. F.J.J. Buytendijk, *Pain*, trans. Eda O'Shiel (London: Hutchinson, 1961), 114.
16. Ibid., 124.
17. Bernhard Waldenfels, *Phenomenology of the Alien: Basic Concepts*, trans. Alexander Kozin and Tanja Stähler (Evanston, IL: Northwestern University Press, 2011), 31.
18. Sigmund Freud, *Inhibitions, Symptoms and Anxiety*, trans. Alix Strachey (London: Hogarth Press, 1936), 169. Similarly Henri Bergson, *Matter and Memory*, trans. Nancy Margaret Paul and W. Scott Palmer (New York: Zone Books, 1991), 56.
19. Freud, *Inhibitions, Symptoms and Anxiety*, 169.
20. Straus, *Vom Sinn der Sinne*, 215.
21. Cf. Christian Grüny, "Zur Logik der Folter," in *Gewalt-Verstehen*, ed. Burkhard Liebsch and Dagmar Mensink (Berlin: Akademie-Verlag, 2003).
22. Kurt Goldstein, *The Organism: A Holistic Approach to Biology Derived from Pathological Data in Man* (New York: Zone Books, 1995), 49.
23. Agustín Serrano de Haro, "Pain Experience and Structures of Attention: A Phenomenological Approach," in *Meanings of Pain*, ed. Simon van Rysewyk (Cham: Springer Verlag, 2016).
24. Lawrence LeShan, "The World of the Patient of Severe Pain of Long Duration," *Journal of Chronic Diseases* 17 (1964): 120.
25. Buytendijk, *Pain*, 57.

26. Geniusias, "Phenomenology of Chronic Pain: De-personalization and Re-personalization," 156.
27. Merleau-Ponty, *Phenomenology of Perception*, 248.
28. Guy Douglas, "Why Pains are not Mental Objects," *Philosophical Studies* 91 (1998): 131.
29. Sartre, *Being and Nothingness*, 332.
30. Abraham Olivier, "When Pains are Mental Objects," *Philosophical Studies* 115 (2003): 48.
31. Friedrich Nietzsche, *The Gay Science: With a Prelude in German Rhymes and an Appendix of Songs*, trans. Josefine Nauckhoff (Cambridge: Cambridge University Press, 2001), 177. Cf. also Christian Grüny, "Vom Nutzen und Nachteil des Schmerzes für das Leben," in *Schmerz als Grenzerfahrung*, ed. Rainer-M.E. Jacobi and Bernhard Marx (Leipzig: Evangelische Verlagsanstalt, 2011).
32. Elaine Scarry, *The Body in Pain: The Making and Unmaking of the World* (Oxford and New York: Oxford University Press, 1985).
33. Ibid., 31.
34. Merleau-Ponty, *Phenomenology of Perception*, 115 (translation modified). Smith translates "schéma corporel" as "body image," which completely misses the point. Cf. Shaun Gallagher, "Body Image and Body Schema: A Conceptual Clarification," *The Journal of Mind and Behavior* 7, no. 4 (Autumn 1986).
35. Drew Leder, *The Absent Body* (Chicago and London: The University of Chicago Press, 1990).
36. Ibid., 26, 53.
37. Cf. Edmund Husserl, *Ideas Pertaining to a Pure Phenomenology and to a Phenomenological Philosophy, Second Book: Studies in the Phenomenology of Constitution*, trans. Richard Rojcewicz and André Schuwer (Dordrecht: Kluwer Academic Publishers, 1989), § 60.
38. Merleau-Ponty, *Phenomenology of Perception*, 473.
39. What it also obscures is the fact that the effortless ease of the lived body relies on an accommodating world, both socially and materially, which presupposes a "normal" body. Cf. for the role of gender Iris Marion Young, "Throwing Like a Girl: A Phenomenology of Feminine Body Comportment Motility and Spatiality," *Human Studies* 3 (1980); for a reflection on disability Vivian Sobchack, " 'Choreography for One, Two, and Three Legs' (A Phenomenological Meditation in Movements)," *Topoi* 24 (2005).
40. Merleau-Ponty, *Phenomenology of Perception*, 107.
41. Ibid., 106–107.
42. Cf. Maurice Merleau-Ponty, *The Visible and the Invisible*, trans. Alphonso Lingis (Evanston, IL: Northwestern University Press, 1968).
43. Leder, *The Absent Body*, 69–99.
44. Cf. Martin Heidegger, *Being and Time*, trans. Joan Stambaugh (Albany: State University of New York Press, 2010), 67–75. Stambaugh's choices of "useful things" for *Zeug* and "handy" for *zuhanden* strike me as being less clumsy and artificial than the old "equipment" and "ready-to-hand."
45. Merleau-Ponty, Phenomenology of Perception, 61.
46. Herbert Plügge, *Der Mensch und sein Leib* (Tübingen: Niemeyer, 1967), 76.
47. Ibid., 63.
48. Husserl, *Ideas Pertaining to a Pure Phenomenology and to a Phenomenological Philosophy*, 42.
49. Cf. Hans Jonas, "Causality and Perception," *The Journal of Philosophy* 47, no. 11 (May 1950).

50. Scarry, *The Body in Pain*, 15.
51. Leder, *The Absent Body*, 87.
52. Merleau-Ponty, *The Visible and the Invisible*, 124.
53. Waldenfels, *Phenomenology of the Alien*, 49–50.
54. Cf. Helmuth Plessner, *Die Stufen des Organischen und der Mensch* (Berlin: de Gruyter, 1975), ch. 7.
55. For the political issues at stake here cf. Darius Rejali, *Torture and Democracy* (Princeton, NJ: Princeton University Press, 2007).
56. Geniusias, "Phenomenology of Chronic Pain: De-personalization and Re-personalization."
57. Olivier, *Being in Pain*, ch. 7.
58. Arthur Kleinman, *The Illness Narratives: Suffering, Healing, and the Human Condition* (New York: Basic Books, 1988), 57.
59. Scarry, *The Body in Pain*, 7.
60. Cf. Merleau-Ponty, *The Visible and the Invisible*, 141.
61. Emmanuel Levinas, *Otherwise than Being, or Beyond Essence*, trans. Alphonso Lingis (Pittsburgh, PA: Duquesne University Press, 1998), 51.

8   Toward a Phenomenology of Fatigue

Katherine J. Morris

This chapter is an initial phenomenological exploration of tiredness, fatigue, chronic fatigue and chronic fatigue syndrome (CFS). It falls into three sections. The first looks at tiredness and fatigue, the second section at fatigue and chronic fatigue (as suffered, *inter alia*, by individuals diagnosed with CFS), and the third at CFS as a biomedical diagnosis.

There are two distinguishable but interrelated questions that I want to address via these investigations. First, it is somewhat curious that phenomenologists have devoted so little attention to such everyday experiences as tiredness and fatigue: both are part of everyday experience—in the case of tiredness, more or less literally *every day*—yet receive little explicit phenomenological attention (unlike, for example, pain).[1] Might doing so lead us to uncover previously neglected dimensions of experience and even to question certain assumptions which some influential phenomenologists have made? Second, there is scope for dialogue between phenomenologists on the one hand and those who are involved in caring for or researching chronically fatigued individuals on the other. What might we as phenomenologists have to learn from them and from individuals who suffer with chronic fatigue or CFS?

Tiredness and Fatigue

The term "tired" is ambiguous; it can mean something like "ready for or in need of sleep" (perhaps "having a felt need for sleep"), but can also mean something like "fed up" or "bored": "I'm tired of you droning on and on about phenomenology" (in this usage, often in the idiom "sick and tired"), as well as "hackneyed" ("a tired cliché"). It is primarily the first usage which will be our focus here, although we should not assume that they are entirely independent. What follows are a few fairly obvious contributions to a phenomenology of tiredness. These remarks mainly occur under headings corresponding to the sorts of categories of experience to which phenomenologists have drawn our attention, although we should never forget that these categories are thoroughly intertwined, nor should we presuppose that the usual categories are the only important ones; I try to work dialectically between these categories and the phenomena with which we are all familiar.

The world: tiredness "can be indicated by objects of the world," for example, by the book which I am trying to read.

> It is with more difficulty that the words are detached from the undifferentiated ground which they constitute; they may tremble, quiver; their meaning may be detached only with effort, the sentences which I have just read twice, three times may be given as "not understood," as "to be re-read."[2]

Objects—e.g. the book—become heavier, distances—e.g. the distance from my study to the bedroom—become longer, inclines—e.g. the stairs—become steeper, and so on. Objects' solicitations become less urgent: the phone doesn't send me leaping out of my chair, a television program that might entice me when I am wide awake loses its appeal, and so on. As it were, the intentional threads that bind me to the world are somewhat slackened. (Here, perhaps, is an overlap between tiredness and boredom.)

The body: even without any part of my body being the explicit object of thetic consciousness, there is in tiredness a kind of background awareness of the body as the correlate of the tired world just sketched. The trembling of the words on the page is the correlate of my eyelids becoming heavy and drooping; the increasing heaviness of the book is the correlate of my arms' finding it more difficult to continue holding it upright; the lengthening of distances and the steepening of inclines are the correlate of body's becoming "heftier," to use Zaner's term.[3]

Fuchs suggests that tiredness involves a "general existential feeling": one of a number of "background feeling states that are characterized by the tacit presence of the body in the experiential field"; these, he claims, include "states of feeling healthy, fresh, strong, or, on the other hand, tired, weak, ill . . . open and alert or indifferent to everything."[4] He doesn't expand on the general existential feeling of tiredness, but words like "lethargy," "lassitude," or possibly "languor" and so on seem more or less to capture it, and we might understand it as the bodily correlate of the "slackening of the intentional threads" referred to earlier.

One might conceptualize tiredness as the body's project to sleep. This bodily project has a complex relationship with one's own projects. My own projects may prevail over my body's, spontaneously—the film was so gripping that I forgot my tiredness—or by "an effort of the will," or by my drinking another cup of coffee or popping a pill; my body's projects may prevail over my own projects—I fall asleep over the pile of essays; or I may simply accede, with perfect contentment, to my body's projects.

Temporality: there is a temporality associated with ordinary tiredness that is distinct from the usual modes of temporality (before/after, past/present/future): a lived diurnality, which we might represent as more like "early/late." To be sure, one can be tired apart from this lived diurnality—e.g. after a long walk—but this type of tiredness is closer to fatigue (which is considered later). Very roughly speaking—such temporality would differ in cultures which have siestas; it is, moreover, disrupted by jet-lag, etc.; and there are individual and age-related differences in sleep cycles—we live the passage of time from morning to evening (from early to late) *in* living distances becoming longer, the body becoming heftier, and so on.

Being-for-others and being-with others: tiredness is often visible by others (in drooping eyelids, slowness of movement and speech, yawning and so on). Ordinary tiredness is readily intelligible by others; indeed tiredness at the end of the day requires no explanation. There is often a kind of "reciprocity," to use Merleau-Ponty's term, or "bodily resonance," to use Fuchs' term, with tiredness: yawning is often contagious, and when we are with someone tired, we often begin to feel tired too, although this reciprocity can manifest itself the other way around: being with a lively companion can wake me up.

Cognition: one doesn't often see phenomenologists talking about cognition. Nonetheless it does seem a crucial aspect of the phenomenology of tiredness that, intertwined with one's movements involving more effort and with the world becoming more difficult, one's thinking becomes slower and involves more effort ("too tired to think straight"), one struggles to find the right words, and so on.

There's a good deal more that could be said, but let's turn to fatigue. The term "fatigue" is also ambiguous, but in the relevant usage may seem to be simply a synonym of "extreme tiredness" (being "tired out" or "dog-tired").[5] I will in fact suggest that, phenomenologically speaking, the distinction between tiredness and fatigue is *not* simply one of degree. Let's look at each of the dimensions above in connection with fatigue.

The world: plausibly, the aforementioned observations on the tired world transfer over to the "fatigued world" with a mere increase in degree: the words quiver *more*, the book becomes *heavier*, distances *longer*, inclines *steeper*, solicitations *less compelling*, etc.

The body: in fatigue too, qualities of the body are the correlate of the qualities of the world. Yet the body's qualities, e.g. its heft, seem not to remain so resolutely in the background; one may, in fatigue, quite explicitly feel one's body as a burden to be hauled up the stairs. One may in such cases (*pace* Sartre) experience a "sensation of effort."[6] Fatigue too, perhaps, involves a "general existential feeling" something like lethargy or lassitude (possibly not languor—and why not?). Fatigue, suggests Bloechl, is "a lapse or diminishment in our capacity to go on."[7] Fatigue "presents itself first as a stiffening, a numbness, a way of curling up into oneself."[8] This "numbness," says Levinas,

> is an impossibility of following through, a constant and increasing lag between being and what it remains attached to, like a hand little

by little letting slip what it is trying to hold on to, letting go even while it tightens its grip. Fatigue is not just the cause of this letting go, it is the slackening itself.[9]

Fatigue is often associated with the body or parts of the body *aching* in a way that tiredness usually isn't. (Perhaps this is what Sartre is referring to when he describes fatigue as becoming "painful.")[10]

Sartre's well-known discussion of fatigue has to do with the interplay between bodily and personal projects, although he would not put it this way: "I start out on a hike with friends. At the end of several hours of walking my fatigue increases. . . . At first I resist and then suddenly I let myself go, I give up, I throw my knapsack down on the side of the road."[11] That I and my companions "suffer" our fatigue differently is down to our different personal projects: I yield my personal project to climb the mountain to my body's project (although for Sartre this is because of my own "fundamental project"), my companion's personal project (because of his fundamental project) prevails over his body's.[12] However, I am inclined to say that whereas the body's project in tiredness is best described as the project to *sleep*, the body's project in fatigue is to *rest*, and perhaps even to obtain various modes of self-care (massaging one's leg muscles, and so on).

Temporality: the temporality of fatigue is importantly different from that of tiredness. Fatigue accumulates over a period of time which differs from case to case—a working week, unaccustomed amounts of exercise—which is somewhat independent from the usual diurnal accumulation of tiredness. It also dissipates according to its own temporality (which, again, differs from case to case) and its dissipation may vary according to the modes of self-care undertaken.[13]

Being-for-others and being-with others: fatigue is likewise often visible by others, primarily in slowness of movement and speech. Fatigue, unlike much ordinary tiredness, does seem to require explanation; but the sorts of explanations that a fatigued person gives for his/her fatigue are readily socially intelligible: "It's been a hard week," "I was up all night finishing my essay," "I have been hiking up this hill for the past three hours." Fatigue too is prone to "reciprocity" or "bodily resonance" at least in one direction: "I'm exhausted just looking at you!" On the other hand, it is less clear that being with someone who is not fatigued can pull one out of fatigue.

Cognition: here, too, fatigue can make thinking and the expression of thoughts slower and more effortful. Again, there is no doubt a good deal more to be said. But let us turn to chronic fatigue.

Fatigue and Chronic Fatigue

Chronic fatigue is commonly associated with a number of illnesses including multiple sclerosis, systemic lupus erythematosus and rheumatoid arthritis; it is not confined to those diagnosed with CFS (on which the next section focuses). In what follows, I assume a broadly similar phenomenology for all forms of chronic fatigue. As the next section shows, however, there are respects in which this is clearly not the case. Moreover, CFS may be "mild," "moderate" or "severe," so that the phenomenology of CFS is unlikely to be uniform.[14] Nonetheless the description that follows should resonate with many sufferers.

Phenomenologically relevant descriptions of chronic fatigue require us to rely on narratives of those who suffer it and the observations of those who work with these people. I draw on a remarkable collection of poetry by CFS sufferers,[15] and on two studies in particular, both by health care professionals and both explicitly couching their observations within a phenomenological framework.[16] The structures we set out in the previous section will facilitate an exploration of the phenomenology of chronic fatigue. We will find that, just as it is tempting but phenomenologically misleading to think of fatigue simply as extreme tiredness, it is also tempting but phenomenologically misleading to think of chronic fatigue simply as long-lasting fatigue.

The world: I quote here from two poems by CFS sufferers, which bring out that the chronically fatigued world is similar to the fatigued world as described earlier.

> She climbs a mountain
> I climb the stairs . . .[17]
>
> Don't ask me if I'm tired
> It makes no sense to me
> Fatigued? Exhausted?
> Doesn't mean a thing
> On my planet
> Gravity's ten times stronger,
> On my planet
> A diver's lead boots are on my feet
> On my planet
> The air's too thin
> On my planet
> A glass of water weighs a ton . . .[18]

It is noteworthy however that the author of the second poem *rejects* the term "fatigued"; in part this surely relates to the temporality of chronic fatigue: these properties of the world are so constant that it is as if the sufferer inhabits a different planet.

The body: "The fatigued body did not feel natural and could not be taken for granted. Some respondents felt betrayed when fatigue came and invaded the body. . . . The body was hard to control and could not be trusted."[19] A sufferer may feel "unsafe within her own body."[20] This is, notably, a theme that one hears closely echoed in narratives from chronic pain patients as well: many people with chronic pain "describe an irrational sense of betrayal, the feeling that faith in their body and the taken-for-granted world have been stolen away."[21] The patient's body may even "become personified as an aversive agent . . . invested with menacing autonomy. 'I think it's against me, that I have an enemy,' Brian said of his body."[22] This can be perhaps understood as an extreme version of the kind of conflict one sometimes feels in tiredness and fatigue between the body's projects and one's own, although it is hard to imagine someone who is fatigued in the ordinary way really personifying his/her body as an aversive agent.

We noted earlier that in ordinary fatigue, the body may be fore-grounded, and it may be felt explicitly as a burden; moreover, it is often associated with parts of the body aching. These characteristics often feature in chronic fatigue too, although again its temporality distinguishes it from ordinary fatigue: "the respondents perceived an increased awareness of their bodies all the time . . . [Sometimes] the body . . . felt heavy and painful as if some hard work had been done."[23]

Sometimes, however, the body in chronic fatigue felt "numbed, dead and not quite awake or even as if parts of it was [*sic*] gone. The body was described by metaphors such as 'a packet' . . . and the legs 'as stumps.' "[24] This seems to be an aspect of chronic fatigue which has no obvious counterpart in ordinary fatigue; it is reminiscent of the type of anosognosia discussed briefly by Merleau-Ponty: "Subjects who systematically ignore their paralysed right hand, and hold out their left hand when asked for their right, nevertheless refer to their paralysed arm as 'a long, cold snake.' "[25] Merleau-Ponty understands anosognosia as a kind of inverse of the phantom limb; in both cases there is a discrepancy between the habit-body and the body at this moment;[26] the world's solicitations call forth a phantom limb, or the limb's present inability to respond to the world's solicitations represses the solicitations and thereby the limb. Although the limbs of chronic fatigue sufferers are not paralyzed, there is, perhaps, a kind of lived analogy.

Temporality: we have already noted that the temporality of fatigue is different from that of tiredness. The temporality of chronic fatigue is, as we might expect, different yet. Indeed medics use the (to my ears trivializing) acronym TATT—"tired all the time"—for chronic fatigue. Thus "fatigue was perceived as nearly always being present."[27] Even when fatigue isn't present, concerns about the possibility of fatigue are set in the structure of the entire day:

> They were thinking about fatigue as occurring at any time, and they always had to take fatigue into account . . . fatigue meant living in a kind of uncertainty. . . . Everything had to be planned in advance and had to take account of the respondents' usual experience of fatigue.

Common daily activities could not be managed without resting in between.[28]

Moreover, insofar as one's life trajectory forms a kind of horizon of temporality for all of one's activities, this too may be impacted by chronic fatigue. In the first place, one's whole life may be experienced as structured around the "before" and "after" of the onset of chronic fatigue. (This point is examined in more depth in the next section.) In the second place, one's relationship to the future may be transformed: "*I don't like using the word career in this particular time in my life because of everything that's happened . . . but there was a time where I . . . had aspirations for a longer career.*"[29]

Ordinary modes of self-care that help to alleviate fatigue and thus to minimize its temporal extent, notably, do not work for chronic fatigue. Individuals with CFS experience, by definition, "overwhelming fatigue that is not improved by rest."[30] Sufferers from chronic fatigue have to learn new regimens of self-care in an attempt to minimize the fatigue or to stop it recurring, often learning from each other at conferences, in online forums, and through publications aimed at sufferers.

Being-for-others and being-with others: like any chronic illness, chronic fatigue can be isolating; sufferers are often unable to engage in activities with others.[31] To add to this source of isolation, "[w]hen fatigued, the respondents were afraid of being mistaken by others to be drunk. Some of them, as a result chose to stay indoors."[32] Thus Johnson reads CFS in part as a "disrupted belonging."[33] (There are further reasons for such isolation in CFS patients discussed in the next section.)

We noted earlier that ordinary fatigue does seem to require explanation; but the sorts of explanations that an ordinarily fatigued person gives for his/her fatigue are socially intelligible. A very striking feature of chronic fatigue is precisely that the sufferer has no readily available socially intelligible explanation for it. (Some of the implications of this for CFS are discussed in the next section.)

Cognition: slowed cognition is so prevalent a feature in CFS that the term "brain fog" has become common parlance amongst sufferers as a "symptom," hence this lovely poem (N.B. ME is another name for CFS; see the next section):

ME Haiku
I was writing a Haiku
On ME
When along came brain-fog and . . .[34]

The difficulty of finding words is likewise brilliantly expressed here:

"*Thingy*," the perpetual understudy, stuttered on a deserted stage;
"*TV remote*" lay bound and gagged in the wings,

> While "the blue thing" (it is black, of course),
> And even "*courgette*" auditioned for the part.[35]

Again, there is undoubtedly a great deal more to be said, but this should at least be sufficient to show that chronic fatigue is not simply long-lasting fatigue. This phenomenological analysis can allow us to begin to understand more deeply why some CFS sufferers resist the term "fatigue" and the label "chronic fatigue syndrome." There are further reasons, as the next section brings out.

Chronic Fatigue and Chronic Fatigue Syndrome

The current WHO's International Classification of Diseases (ICD) includes the following clinical information for CFS:

- A condition lasting for more than 6 months in which a person feels tired most of the time and may have trouble concentrating and carrying out daily activities. Other symptoms include sore throat, fever, muscle weakness, headache, and joint pain.
- . . . Symptoms are not caused by ongoing exertion; are not relieved by rest; and result in a substantial reduction of previous levels of occupational, educational, social, or personal activities . . .
- . . . The etiology of CFS may be viral or immunologic. Neurasthenia and fibromyalgia may represent related disorders. Also known as myalgic encephalomyelitis [ME].
- . . . Since other illnesses can cause similar symptoms, CFS is hard to diagnose. No one knows what causes CFSs. It is most common in women in their 40s and 50s, but anyone can have it. It can last for years. There is no cure for CFS, so the goal of treatment is to improve symptoms.[36]

The term "chronic fatigue syndrome" has a complex history and is hugely contested.[37] The nineteenth-century label "neurasthenia," for which the treatment of choice was Weir Mitchell's famous rest cure, had a similar clinical profile to modern CFS; various "fatigue epidemics" in the twentieth century were connected with viruses, including polio in the 1930s and the Epstein-Barr virus (which causes mononucleosis) in the 1960s, resulting in one label still sometimes in use, "post-viral fatigue syndrome"; the term "myalgic encephalomyelitis" (ME) emerged in the 1950s; "chronic fatigue syndrome" as well as "chronic fatigue and immune deficiency syndrome" or CFIDS emerged in the 1980s; and then there was the disparaging (and notorious) term "yuppie flu," described by *Newsweek* as "a fashionable form of hypochondria."[38] Today many sufferers, and organizations aimed at them, use the clumsy compound "CFS/ME" or "ME/CFS," although Americans tend to prefer "CFIDS."[39] "Thus the configuration of symptoms currently best known as chronic fatigue syndrome has been constructed and reconstructed in popular and professional discourse over time."[40]

This fluctuating terminology, however, is the froth on a sea of more substantive controversies. There was in Britain a Private Member's Bill in 1987 to get ME/CFS recognized by Parliament; as a consequence, a Task Force was initiated which reported in 1994, and this report validated the condition. The government response to this report was set out in the Royal Colleges' *Report on Chronic Fatigue Syndrome*, published in 1996. This document also recognized the condition as a valid illness.[41] However, it spoke of the condition as "existing in the grey area between mind and body."[42] And some patients dislike the term "chronic fatigue syndrome" precisely because it seems to imply "psychiatric causation."[43] The ICD classifies CFS/ME as a neurological disorder (and hence not as psychiatric). However, during the latest revision of DSM—the Diagnostic and Statistical Manual of Mental Disorders, lovingly known as "the psychiatrists' bible"—the CFIDS Association of America was moved to respond to the DSM-5 Task Force's open request for input on proposed changes. Their worry concerned the Task Force's proposed rubric of "complex somatic symptom disorder" (CSSD): it looked as though many CFS patients would meet enough of the criteria for CSSD to be diagnosed with this mental disorder. It concludes by asking the Task Force to abandon the proposed classification, to avoid "increased harm to patients through confusion with other conditions or attaching further stigma."[44]

Several things emerge from this brief look at the controversy: in particular, several *cultural assumptions*. (i) Distress, at least long-lasting, life-impacting distress, must be interpreted as a disorder (i.e. medicalized). (ii) Disorders ought to be categorizable as either "mental" or "physical." (iii) So-called mental disorders carry stigma. (iv) Mental disorders are in some sense "not real."[45] We might upon reflection wish to add this: (v) disorders that aren't neatly categorizable as either mental or physical likewise carry stigma.[46] That these are cultural *assumptions* is shown by the fact that within Chinese culture (to generalize vastly), these assumptions are not made; and within Chinese medicine, CFS would be understood as a depletion of *qi*, which sits outside the duality of "mind" and "body."[47]

I want to suggest that these cultural assumptions, insofar as they are shared by sufferers, family, friends, colleagues and health care providers, make the lived experience of CFS importantly different from that of many other conditions that involve chronic fatigue (but also make it similar, once again, to chronic pain).[48] In particular, they (together with certain features of the syndrome itself) impact upon temporality and upon being-for-others and being-with-others. To examine this, we will draw on sufferers' narratives as well as a different set of studies, this time by social scientists.[49]

Temporality: we noted earlier that the lives of those suffering with chronic fatigue may be seen as structured around the "before" and "after" of its onset. This, however, is apt to be true of *anyone* with a major, especially a chronic, illness: it "disrupts the trajectory of a person's biography, undermining his/her identity, self-reliance and social relationships."[50] What is particular to CFS (as well as chronic pain) is the long period of searching for a diagnosis, which may aptly be characterized as a *quest*: "the quest for an identifiable illness to give meaning to their experiences."[51] Many people "experience a delay in achieving a diagnosis of CFS."[52] There are multiple reasons for this; one is that a diagnosis of CFS needs to be *negotiated* with the medical profession. Many sufferers receive diagnoses of other conditions—especially depression, but also, e.g. pre-menstrual tension or the menopause—and some medics are skeptical about the existence of CFS.[53] "Their attempts at obtaining a diagnosis were constantly rejected, their symptoms were dismissed and disbelieved, and they were often labelled as bored housewives or depressed adolescents."[54] According to one, the doctor said "all right cards on the table now what's wrong with you? Something wrong with your marriage?"[55] "Hence the search for a diagnosis was often a lengthy ordeal with numerous unfruitful meetings, which often led to difficult relationships between the participants and their care providers."[56] Many sufferers end up going to multiple doctors: "the need for a technical and rational representation of the 'disease' often entails changing specialist on a number of occasions."[57] Moreover, "a diagnosis of CFS/ME was not the end of the journey and, in fact, may have only been the beginning in the search for treatment and restoration of health status."[58]

It is easy enough to see why the diagnosis of a condition with no clear biological markers might take time. It is also easy enough to see why it would be desirable for CFS sufferers to have a diagnosis of some sort; it, for example, may give them access to benefits, be they tangible (e.g. disability benefits[59]) or intangible (the right to perform the "sick role"[60]). But what is a "technical and rational representation of the 'disease,'" and why exactly are sufferers so resistant to diagnoses of, for example, depression or PMT?[61] Moreover, although receiving the label "CFS/ME" gives sufferers some relief "despite the uncertainty and stigma surrounding the label,"[62] the quest often continues to try to identify CFS as an unambiguously *physical* disorder.[63] The desire for CFS to be a physical and not a mental disorder is evident in the popularity of the subtitle of Myhill's 2014 book: "Mitochondria, not hypochondria."[64] Are the cultural assumptions identified earlier not at work here?

Being-for-others and being-with-others: the concept which figures most prominently in social scientists' analyses of the intersubjective experience of CFS sufferers is "delegitimation" (closely linked with stigma), defined as "the experience of having one's perceptions of an illness systematically disconfirmed."[65] Ware identifies two types of delegitimizing encounters which "appeared regularly in interviewees' reports of their experience with chronic fatigue syndrome."[66]

The first type "stems from the apparent insignificance of the symptoms,"[67] "and the inconsistency or variability of symptoms, in that the person appeared 'better' on some days than on others."[68] Perceptions of the trivialization of symptoms by others converge for sufferers in the thematic phrase: "You're tired? We're all tired! So what!"[69] "I was just left with a feeling 'Oh Irene's made a fuss again' . . . that's the sort of attitude they have toward ME people."[70] "If you have cancer, you can tell your friends you have cancer and your friends understand. You cannot tell your friends you are tired. What are they going to say? 'I'm tired too!'"[71] One consequence is that sufferers may get little sympathy (one CFS poem is entitled "Can't find sympathy");[72] "they think 'Oh, they're at it, they're just making a meal of this' or whatever."[73] Indeed the absence of sympathy may extend as far as outright skepticism; and again, this experience is echoed by chronic pain sufferers: "[i]f there is a single experience shared by virtually all chronic pain patients it is that at some point those around them . . . come to question the authenticity of the patient's experience of pain."[74] Worse yet, CFS sufferers "are subject to accusations of lying about or exaggerating their condition as a means to escape mundane responsibilities."[75] This is a further source for the isolation which many chronically ill people experience: "Having a shadow cast over their moral characters can then result in apprehension about disclosing their condition to others and often secrecy."[76]

The second type "is embodied in physicians' definition of the illness as psychosomatic—'all in your head,'" often on the grounds that "no observable evidence of disease in the form of clinical signs or laboratory findings can be found."[77] The connection made by both sufferers and medics between "no observable evidence of disease in the form of clinical signs or laboratory findings" and "all in your head," i.e. "not real," is likewise found with chronic pain syndrome: when sufferers are told that "that there was no organic basis for [their] problems,"[78] some interpret the diagnosis as suggesting that that they are somehow to blame for their suffering or that they're "crazy."[79] Others take it that what is being said is "that the pain is not real, that it's in your mind, in your head, it's not a real pain,"[80] thus "reproducing dualistic thinking and sharing the stigmatized status of mental illness as a disability we 'bring on ourselves.'"[81]

> In fearing that their illness might, after all, be psychosomatic, self-doubters confront the possibility of psychological disorder and the stigma it entails . . . [thus] adding the burden of a stigmatized identity to that of living with a chronic illness that is severely debilitating, basically untreatable, and of questionable authenticity in the eyes of others.[82]

So the particular features of CFS, together with the cultural assumptions identified earlier, impact dramatically both on sufferers' lived temporality and on their being-with-others and being-for-others, and, importantly, add to their suffering: "[o]f the various forms of suffering that experiences of delegitimation can engender, none was as devastating for this group as the humiliation that resulted from having their subjective perceptions and sensations of illness either trivialized or dismissed as psychosomatic."[83]

Some Concluding Remarks

A comment: there are delicate issues about the terms "normality" and "abnormality" here. It would be possible, though uncharitable, to read Merleau-Ponty as sharply distinguishing a kind of idealized "normal" experience—the experience of an inconceivable human being who, for instance, never gets tired—and that of the clearly "abnormal" (e.g. an individual with a phantom limb).[84] There are a couple of points to be made here. First, if fatigue was simply extreme tiredness and chronic fatigue were simply long-lasting fatigue, then the different between normal and abnormal here would seem simply to be a matter of degree. Yet we have seen that, phenomenologically speaking, neither of these is the case, so the conclusion does not follow: chronic fatigue really is qualitatively different from ordinary fatigue. But, second, we should not be timid about using the word "abnormal": norms need not be either social norms or statistical norms, and clearly are not in the ways in which Merleau-Ponty invokes them. Rather, we must see them as at least akin to Canguilhem's *vital* norms.[85]

I will suggest some tentative conclusions: in the first place, as I said at the beginning, it does seem strange that phenomenologists have devoted so little attention to such everyday experiences as tiredness and fatigue. Our initial exploration revealed some important dimensions of lived experience that could be further explored: for example, different modes of temporality such as "early/late"; and the potential importance of a phenomenology of cognition. It also provided materials for calling into question certain assumptions sometimes made by phenomenologists.

First, it seemed entirely natural to conceptualize tiredness as "the body's project to sleep," and fatigue as "the body's project to rest"; yet not all phenomenologists of the body are able to accommodate such a conception. For example, Sartre conceives of the body as the facticity of the for-itself,[86] whereas projects by their very nature involve transcendence. Merleau-Ponty, by contrast, identifies a form of intentionality—what he, following Husserl, calls "operative intentionality"[87]—which is the body's own. However, for Merleau-Ponty, my body's projects seem always to be in the service of my own: "I want to go over there, and here I am . . . I look at the goal, I am drawn by it, and the bodily apparatus does what must be done in order for me to be there."[88] Yet in tiredness and fatigue, it seemed natural to see at least a potential conflict between the body's projects and my own; indeed, in chronic fatigue, the body might be experienced as a positively aversive agent.

Second, Leder's (1990) distinction between the recessive and the "dys-appearing" body is rightly celebrated; pain—at least a sudden acute pain—is a paradigm case in which the painful part of the body moves from background to foreground. It would be possible, though unchari-table, to read him as suggesting a *sharp* distinction between recession and dys-appearance. And neither tiredness nor fatigue seem to allow this, in part because they are both cumulative, and in part because it sometimes seems to be the entire body which (for example) becomes bur-densome. Similar considerations might lead to the suggestion that the distinction between non-thetic and thetic awareness is not always sharp, and likewise the distinction between the body-for-itself and the body-in-the-midst-of-the-world, which Sartre sees as existing on "two incom-mensurable levels."[89]

One might go further. For Sartre, "[t]here can be a free for-itself only as engaged in a resisting world."[90] Levinas takes him to task for stopping here: "the conception of action which is being presupposed in this image of a struggle with matter" is simply taken for granted; we need to start with "the instant of effort and its internal dialectics": "[e]ffort lurches out of fatigue and falls back into fatigue."[91] The possibility of fatigue is essential to the possibility of *action*, we might say.

In the second place, the second section showed how a phenomenological approach could illuminate chronic fatigue, and in particular, how it could justify resistance to the label "chronic fatigue," which makes fatigue and chronic fatigue sound too much alike. This, however, is a point which health care professionals have already understood. What has been less understood by phenomenologists is the ways in which they themselves can learn from health care professionals and social scientists; this is what I tried to bring out in the third section. In particular, it revealed the ways in which cultural assumptions can impact upon lived temporality and on being-with-others and being-for-others.[92,93] Phenomenologists have long accepted the princi-ple that the body is always already cultured; they have not always explored the particular ways in which cultural assumptions are bodily lived. More strongly, phenomenologists have long been critical of so-called Cartesian-ism in philosophy; if a culture's assumptions are in some sense Cartesian, might phenomenology not contribute to a cultural critique?

Notes

1. An interesting exception is Levinas, whose analysis I touch on from time to time; see Emmanuel Levinas, *Existence and Existents*, trans. Alphonso Lin-gis (Dordrecht, Boston and London: Kluwer Academic Publishers, 1978).

2. Jean-Paul Sartre, *Being and Nothingness*, trans. Hazel E. Barnes (London: Routledge, 1986), 332. This passage in fact concerns pain in the eyes, although it seems equally applicable to tiredness.
3. Richard M. Zaner, *The Context of Self* (Athens, OH: Ohio University Press, 1981), 55 ff.
4. Thomas Fuchs, "The Phenomenology of Affectivity," in *The Oxford Handbook of Philosophy and Psychiatry*, eds. K.W.M. Fulford, Martin Davies, Richard G.T. Gipps, George Graham, John Z. Sadler, Giovanni Stanghellini and Tim Thornton (Oxford: Oxford University Press, 2013), 614–15.
5. Somewhat disconcertingly for philosophers, googling "phenomenology of fatigue" reveals such titles as Pietro Paolo Milella, "Nature and Phenomenology of Fatigue," in *His Fatigue and Corrosion in Metals* (Milano: Springer, 2013).
6. Sartre, B&N, 304. See Michael Gillan Peckitt, "Resisting Sartrean Pain," in *Sartre on the Body*, ed. Katherine J. Morris (Basingstoke and New York: Palgrave Macmillan, 2010) for a related critique of Sartre on this point. The notion of effort (although not necessarily "sensations of effort") is central to Levinas' analysis of fatigue (*Existence*).
7. Jeffrey Bloechl, "The Difficulty of Being: A Partial Reading of E. Levinas, *De l'existence à l'existant*," *European Journal of Psychotherapy, Counselling and Health* 7, nos. 1–2 (March–June 2005), 78.
8. Levinas, *Existence*, 16.
9. Ibid.
10. Sartre, B&N, 453.
11. Ibid.
12. Sartre goes further: his hiking companion "*loves* his fatigue; he gives himself up to it as to a bath; it appears to him in some way as the privileged instrument for discovering the world which surrounds him" (B&N, 455).
13. Running magazines often advise both on the time typically required for recovery from a marathon and on techniques of self-care to facilitate such recovery. E.g. www.runnersworld.com/for-beginners-only/when-to-return-to-running-after-a-marathon
14. See the "Information for the public" provided by NICE (the (UK) National Institute for Health and Clinical Excellence) at www.nice.org.uk/guidance/cg53/ifp/chapter/what-is-cfsme
15. OMEGA, *Poetry from the Bed* (Oxfordshire Myalgic Encephalomyelitis Group for Action, 2012).
16. Gulvi Flensner, Anna-Christina Ek, and Olle Soderhamn, "Lived Experience of MS-Related Fatigue—A Phenomenological Interview Study," *International Journal of Nursing Studies* 40 (2003)—the authors are health care professionals working with multiple sclerosis sufferers—and Anne Johnson, Exploring the Experiences and Occupations of Men with Chronic Fatigue Syndrome/Myalgic Encephalomyelitis (CFS/ME) Using a Gadamerian Interpretive Phenomenological Framework," PhD diss., University of the West of England, 2017—the author is an occupational therapist working with CFS sufferers.
17. OMEGA, 5 (Lucy Stone).
18. OMEGA, 49 (Lynda Holland).
19. Flensner, Ek and Soderhamn, "Lived Experience," 711.
20. M.A. Arroll and V. Senior, "Individuals' Experience of Chronic Fatigue Syndrome/ Myalgic Encephalomyelitis: An Interpretative Phenomenological Analysis," *Psychology and Health* 23, no. 4 (2008), 12–13.
21. Byron J. Good, *Medicine, Rationality, and Experience: An Anthropological Perspective* (Cambridge: Cambridge University Press, 1994), 127.

22. Ibid., 125.
23. Flensner, Ek and Soderhamn, "Lived Experience," 711.
24. Ibid.
25. Maurice Merleau-Ponty, *Phenomenology of Perception*, 88/79. All page references cite the two English translations in chronological order: Trans. Colin Smith (London and New York: Routledge, 2002), trans. Donald Landes (Routledge: London, 2011).
26. Merleau-Ponty, PofP, 95/84.
27. Flensner, Ek and Soderhamn, "Lived Experience," 711.
28. Ibid.
29. One of Johnson's interviewees, quoted in Johnson, *Exploring*, 207, italics hers.
30. From the Centers for Disease Control and Prevention website, www.cdc.gov/me-cfs/about/index.html
31. See the pair of very moving paintings by CFS sufferer Charlotte Pitcher in the April 2010 OMEGA Newsletter entitled "Isolation 1" and "Isolation 2."
32. Flensner, Ek and Soderhamn, "Lived Experience," 711.
33. Johnson, *Exploring*.
34. OMEGA, 13 (Sandra Fardon Fox). N.B. this is the complete poem.
35. OMEGA, 21 (Ian Seed).
36. This appears to be the latest version, and distinguishes "CFS not otherwise specified" from post-viral fatigue syndrome; for present purposes I need not enter into the details, www.icd10data.com/ICD10CM/Codes/R00-R99/R50-R69/R53-/R53.82
37. See Diane L. Cox, *Occupational Therapy and Chronic Fatigue Syndrome* (London and Philadelphia: Whurr Publishers, 2000), also Norma C. Ware, "Suffering and the Social Construction of Illness: The Delegitimation of Illness Experience in Chronic Fatigue Syndrome," *Medical Anthropology Quarterly* New Series 6, no. 4 (December 1992), 348–9.
38. Newsweek staff, *Chronic Fatigue Syndrome*. www.newsweek.com/chronic-fatigue-syndrome-205712
39. See also the 1998 Research Paper 98/107 for the House of Commons on CFS/ME, available at http://researchbriefings.files.parliament.uk/documents/RP98-107/RP98-107.pdf
40. Ware, "Suffering," 349.
41. It also recommended ceasing to use the term "myalgic encephalomyelitis" or ME as it implies a specific brain pathology that is not found in these patients. The term continues to be used, however.
42. Lesley Cooper, "Myalgic Encephalomyelitis and the Medical Encounter," *Sociology of Health and Illness* 19, no. 2 (1997), outlines this bit of the history (187).
43. Alex Sleator, Chronic Fatigue Syndrome/ME. http://researchbriefings.files.parliament.uk/documents/RP98-107/RP98-107.pdf
44. K. Kimberly McCleary, *Response to a Request from the DSM-5 Task Force for Input on Proposed Changes* (Charlotte, NC: The CFIDS Association of America, 2010). Available online at https://dxrevisionwatch.files.wordpress.com/2010/04/dsm5-statement.pdf
45. The Royal College report on chronic fatigue refers to disagreements "on such questions as whether the syndrome or some form of it exists, whether it is 'real' or 'organic' or 'merely' psychological."
46. Our cultural intolerance for ambiguity might also be identified in the stigma that continues to be attached to non-binary gender.
47. See, e.g. Judith Farquhar, *Knowing Practice: The Clinical Encounter of Chinese Medicine* (New York and Abingdon: Routledge, 1994).

48. Ware, "Suffering," 357 makes this point. See Katherine J. Morris, "Chronic Pain in Phenomenological/Anthropological Perspective," in *The Phenomenology of Embodied Subjectivity*, eds. Rasmus T. Jansen and Dermot Moran (Cham and Heidelberg: Springer Verlag, 2013), on phenomenological reflections on chronic pain and chronic pain syndrome, which, notably, does have a DSM classification. Arguably some of these remarks would apply equally to other chronically fatiguing conditions such as MS prior to its identification in the nineteenth century as a distinct neurological condition.

49. For example, medical anthropologists such as Ware, "Suffering," sociologists such as Cooper, "Medical Encounter," and Cristina Lonardi, "The Passing Dilemma in Socially Invisible Diseases: Narratives on Chronic Headache," *Social Science and Medicine* 65 (2007), and health psychologists such as Adele Dickson, Christina Knussen and Paul Flowers, "Stigma and the Delegitimation Experience: An Interpretative Phenomenological Analysis of People Living with Chronic Fatigue Syndrome," *Psychology and Health* 22, no. 7 (2007), also "'That Was My Old Life: It's Almost Like a Past-Life Now': Identity Crisis, Loss and Adjustment Amongst People Living with Chronic Fatigue Syndrome," *Psychology and Health* 23, no. 4 (2008).

50. Lonardi, "Passing," 1619; see also Dickson, Knussen and Flowers, "'That Was My Old Life.'"

51. Arroll and Senior, "Individuals' Experience," 14.

52. Dickson, Knussen and Flowers, "Stigma," 862.

53. Ibid., 856.

54. Cooper, "Medical Encounter." 190.

55. Quoted in Cooper, "Medical Encounter," 198.

56. Arroll and Senior, "Individuals' Experience," 20.

57. Lonardi, "Passing," 1623.

58. Arroll and Senior, "Individuals' Experience," 24.

59. See, e.g. www.disability-benefits-help.org/disabling-conditions/chronic-fatigue-syndrome; the ME Association's submission to the independent Work Capability Assessment Benefit Review (September 2010), www.meassociation.org.uk/2010/09/me-association-submission-to-the-independent-work-capability-assessment-benefit-review/

60. See e.g. Cooper, "Medical Encounter," 196 and 202–3, Dickson, Knussen and Flowers, "Stigma," 853.

61. Dickson, Knussen and Flowers, "Stigma," 856f. "Psychiatrists argue that attributing an aetiological role to psycho-social factors does not deny legitimate access to the sick role, whilst sufferers argue vehemently in the self help journals and in the press against being labelled as mentally ill, even if that does mean that they are given an entry points into the sick role" (Cooper, "Medical encounter," 189).

62. Cooper, "Medical Encounter," 196. She continues: "[w]hen symptoms are eventually labelled as Myalgic Encephalomyelitis or Chronic Fatigue Syndrome this is often done in a manner that nonetheless denies the validity of the label."

63. Cooper, "Medical Encounter," 196.

64. Sarah Myhill, *Diagnosis and Treatment of Chronic Fatigue Syndrome: Mitochondria, Not Hypochondria* (London: Hammersmith, 2014).

65. Arthur Kleinman, "Pain and Resistance: The Delegitimation and Relegitimation of Local Worlds," in *Pain as Human Experience: An Anthropological Perspective*, eds. Mary-Jo DelVecchio Good, Paul E. Brodwin, Byron J. Good and Arthur Kleinman (Berkeley: University of California Press (paperback edition), 1994).

66. Ware, "Suffering," 350. Dickson, Knussen and Flowers ("Stigma") structure their article around "Negotiating a diagnosis of CFS" (856 ff.) and "Negotiating CFS with loved ones" (858 ff.); very broadly these negotiation issues correspond to Ware's second and first type of delegitimation encounter, respectively.
67. Ware, "Suffering," 350.
68. Dickson, Knussen and Flowers, "Stigma," 862; cf. Lonardi, "Passing," 1625.
69. Ware, "Suffering," 350.
70. Quoted in Cooper, "Medical Encounter," 197.
71. Sufferer quoted in Ware, "Suffering," 350.
72. OMEGA, 54 (Linda Angeletta).
73. CFS sufferer quoted in Dickson, Knussen and Flowers, "Stigma," 859.
74. Arthur Kleinman, *The Illness Narratives: Suffering, Healing, and the Human Condition* (New York: Basic Books, 1988), 57.
75. Dickson, Knussen and Flowers, "Stigma," 862; cf. Lonardi, "Passing," 1625.
76. Ibid.
77. Ware, "Suffering," 350.
78. Linda C. Garro, "Chronic Illness and the Construction of Narratives," in *Pain as Human Experience*, eds. DelVecchio Good, Brodwin Good and Arthur Kleinman, 112.
79. Ibid., 104. The notion that we are somehow responsible for our own mental problems may be another cultural relic of Cartesianism; thanks to Rupert Read for making me think about this.
80. Quoted in J.E. Jackson, in the significantly titled "'After a While No One Believes You': Real and Unreal Pain," in *Pain as Human Experience*, eds. DelVecchio Good, Good Brodwin, and Arthur Kleinman, 1994, 165, no. 30.
81. Ware, "Suffering," 356.
82. Ibid., 352.
83. Ibid., 353–4.
84. Some of these assumptions were highlighted in Katherine J. Morris, "The Phenomenology of Clumsiness," in *Sartre on the Body*, ed. Morris, 2010.
85. As was argued in Katherine J. Morris, *Starting with Merleau-Ponty* (London and New York: Continuum, 2012), Coda to Ch. 3.
86. At least in B&N; see 306 ff.
87. Merleau-Ponty, PofP, xx, lxxxii.
88. Maurice Merleau-Ponty, *Signs*, trans. R.C. McCleary (Evanston, IL: Northwestern University Press, 1964), 66.
89. Sartre, B&N, 304.
90. Ibid., 483.
91. Levinas, *Existence*, 17.
92. "This was also the aim of Morris" "Chronic Pain."
93. There is much more to be said: I have, for example, said nothing here about gender and the cultural assumptions around gender, but it is noteworthy both that the majority of those diagnosed with CFS are women, and that women under patriarchy have long suffered from delegitimation experiences. Fricker's notion of hermeneutic injustice, as developed in Miranda Fricker, *Epistemic Injustice: Power and the Ethics of Knowing* (Oxford: Oxford University Press, 2007), links to this.

Section III

Recovery and Life's Margins

9   Suffering's Double Disclosure and the Normality of Experience

James McGuirk

Introduction

My intention in the present chapter is to offer some reflections on approaches to the phenomenon of suffering in phenomenological discourses, as well as the implications of these approaches for the ways we understand the structure of normality in experience. Analyses of suffering, I will show, are doubly disclosive in the sense of proffering insight into the nature of certain pathologies, illnesses, and injuries as well as delineating some of the most fundamental aspects of everyday experience from which suffering alienates us. Such analyses are, in other words, disclosive of suffering itself, as well as offering a *via negativa* to understanding aspects of ordinary world experience.[1] But we will also see that the doubleness of this disclosure is more dialectic than binary as a result of the possibility for sufferers to reclaim or reconstitute normality in the face of pathology or illness.

The article is divided into three parts. The first two parts deal with different ways in which the relationship between suffering and normality have been conceptualized in different phenomenological discourses. The first part deals with the contribution made by phenomenology to psychopathology and specifically to the articulation of the lived experience of mental illness. Work in phenomenological psychopathology has been clinically fruitful in articulating the first-person presentation of specific pathologies, but has also proved important in illuminating subjective and intersubjective lived experience in a more global way. It does so by allowing close analyses of pathological experience to throw light on aspects of experience we take for granted pre-reflectively, but which we are often opaque for us from the point of view of ordinary reflection. In the second part, I take up work done on the phenomenology of illness whose focus is less on pathological and normal subjective world experience as such, but on the capacity of the subject to live with suffering and loss in the context of illness. One of the central focal areas of this work has been to draw attention to the way a sense of the normal can be reconstituted in spite of the loss of capacity that accompanies illness and suffering. The point here is not to conceive of suffering in terms of alienation from the normal, but as something woven into ordinary experience in a way that challenges a clear-cut delineation, not only of the categories of health and illness, but of normal and abnormal. In the third and final part, I attempt to reconcile these positions by arguing that while we should avoid thinking of normal and abnormal in absolute terms, we must be equally careful to avoid conceiving of the normal as endlessly malleable through modification or reconceptualization. The normal is not a purely objective criteria imposed from without, but nor is it wholly emergent from within the subject's capacity to re-orient itself in the world. Both the normal and its felt loss in suffering, are comprehensible only in the inter-face between subject, world and others, in which each moment exerts a constraining force on the others. As such, the realization of the normal is both dynamic and limited.

The Normal and the Abnormal: Perspectives from Phenomenological Psychopathology

The first of the approaches to which we turn our attention is one con-cerned with psychopathology and the implications of articulations of the lived experience of the sufferer for understanding subjectivity and intersubjectivity generally. While Husserl and Merleau-Ponty delib-erately developed their phenomenological research in dialogue with empirical psychology and psychiatry respectively, phenomenological psychopathology can also point to a long history as a discrete area of inquiry emerging to a great extent from Karl Jaspers's famous two-volume *General Psychopathology* from 1913.[2] *General Psychopa-thology* is not exclusively a work in the phenomenological tradition, and indeed Jaspers is often skeptical about what a phenomenological approach can contribute to pathologies such as schizophrenia, but the work is nevertheless important for its attempt to anchor our understand-ing of mental illness in the context of the human experience of life as a meaningful unity. And given his commitment to the subject matter of psychopathology as "actual conscious psychic events"[3] and "the intentional relationship of the I to what confronts it,"[4] it is arguable that the spirit of Jaspers' work is more phenomenological than even his own comments on phenomenology would suggest. The *General Psy-chopathology* has also inspired a great many thinkers including Eugene Minkowski, Ludwig Binswanger, Kimura Bin and, more recently, Josef Parnas, Louis Sass and Thomas Fuchs, all of whom have made sig-nificant contributions to psychopathology by using phenomenological resources to closely describe the lived experience of pathologies such as schizophrenia, depression and dementia.

Despite the volume of work done from within this paradigm, it would be wrong to suggest that the phenomenological approach has ever been anything more than marginal within psychiatry. Mainstream psychiatry and psychopathology tend to be dominated by the so-called biomedical paradigm, with its materialist orientation and its emphasis on the bio-chemical etiology of mental illness.[5] This approach has dominated for several reasons, one of which is that it has simply been effective in the analysis and treatment of mental disorders. However, what it neglects, and what phenomenological psychopathology specifically offers, is attention to the lived experience of mental illness. The medical paradigm tends to thematically ignore the disturbance that pathology entails in the life-world of a living subject. It ignores, in other words, the *suffering* of the disorder in the context of a life. Let me be clear here. I do not mean to say that proponents of the biomedical approach are blind or indifferent to the suffering of patients with depression, schizophrenia or any other pathology. Psychiatrists are human beings and they see first-hand the devastation that mental disorder can bring about. The point is rather that the paradigm itself is set up not to take suffering into account when investigating, explaining and treating pathology. That the patient suffers is not in doubt. What is in doubt is whether the suffering *itself* can tell us anything clinically interesting. The question is whether an articulation of the sense of suffering can contribute to our understanding of what the pathology is or if suffering is merely a by-product of pathology properly understood.

It is precisely on this point that advocates of the phenomenological approach differ. They argue that articulating the nature of the first-person suffering of the patient is not only clinically helpful, but necessary if we are to have any hope of understanding or treating disorders. It is important to know why they make this claim. The phenomenologists' focus on the suffering of pathology is not motivated by patient advocacy or on an attempt to supplement the clinical with a focus on care. The phenomenological description is not an addendum to the bio-chemical. It is understood, rather, as necessary to the development of a clinical picture of the disorder by giving as faithful a description of the lived experience of the disorder as possible. As Jaspers says,

> The first step towards knowledge of the psyche is the selection, delimitation, differentiation and description of *particular phenomena of experience* which then, through the use of the allotted term, become defined and capable of identification time and again.[6]

For Jaspers, this is simply good science. If we are going to be able to classify pathology in ways that are nuanced and precise, it is important that we know as much as possible about the thing we are trying to understand. In other words, while it is necessary and desirable to articulate pathology in terms of somatic etiology, its meaning as pathology only makes sense with reference to the touchstone of experience.[7]

Jaspers's aforementioned methodological commitment, combined with his admonition to "discount the theoretical prejudices ever present in our minds and train ourselves to *pure appreciation of the facts*"[8] is very much in the spirit of Husserl's stated aim to make phenomenology an "eidetic science of consciousness" through which the *a priori* of the contents of experience are brought to reflective presence. Within psychopathology, the approach has been used to articulate the subjective expression of a range of pathologies. One might think here of Thomas Fuchs's articulation of depression in terms of a corporealization of the body in which the latter, rather than opening "space as a realm of possibilities, affordances, and goals for action . . . stands in the way as an obstacle."[9] Or the same author's description of the felt disembodiment of mind in schizophrenia in which the sufferers sense of ownership (SO) of their actions becomes dislocated from their sense of agency (SA).[10] Another researcher, Louis Sass, has emphasized the tendency of schizophrenics to exhibit hyper-reflexivity, which is to say that psychophysical agency, which ordinarily recedes into the background of attentional focus, protrudes in a way that makes engagement with the world and with others difficult.[11] In the background to all of these descriptions, lies a commitment to the importance of intentionality in understanding (pathological) experience. Husserl's idea of intentionality—that consciousness, by definition, transcends toward the world while the world, by definition, manifests itself as meaningful to consciousness—was ground-breaking within the phenomenology of mind, though it states nothing more than what we take for granted in our ordinary pre-reflective experience. Our experience of ourselves, the world, and others is enacted in the co-belonging of consciousness and world in ways that are generally established, uncontroversial, and unproblematic. Yet, it is precisely these established structures that become strained for the schizophrenic. Hyperreflexivity denotes the obtrusion of the tacit dimension of experience into the forefront of attention in a way that is seriously debilitating. As Sass puts it, "the perfectly normal sensations implicit in ongoing experience and action [are] now experienced in the perfectly abnormal condition of hyperreflexivity and altered self-affection."[12]

These examples testify to the great potential of phenomenology in articulating the nature and structure of the experience of pathology, in ways that contribute enormously to our understanding and classification of these pathologies.

But the focus on pathology is disclosive in another way too, which has been at least as popular among phenomenologists. Articulating the experience of the schizophrenic not only helps us to understand schizophrenia, but also clarifies aspects of normal world experience that are often so obvious and tacit that we fail to attend to them in our normal self-apprehension. Contrasting the value of thought experiments with the

use of empirical case studies in reflection on the nature of lived experience, Dan Zahavi and Josef Parnas put the point as follows:

> If we are looking for phenomena that can shake our ingrained
> assumptions, and force us to . . . revise . . . our habitual way of think-
> ing, there is no need to get lost in farfetched and unreliable fantasies.
> All we have to do is turn to psychopathology (along with neurology,
> developmental psychology, ethnology etc.) since all of these disci-
> plines present us with rich sources of challenging material.[13]

Pathological experience shows itself as pathological against the background of aspects of experience that are so foundational that we fail to
notice them. But, as Heidegger famously pointed out, it is often only in
the kind of breakdown (including the breakdowns involved in pathology)
of the normal progression of things that the ordinary aspects of experience can become reflectively available to us.[14] Thus, while the pathological is experienced *as pathological* because of the ordinary nature of
comportment which we take for granted, ordinary comportment only
comes to our attention when something has gone wrong. The best-
known exploitation of this insight in the phenomenological literature is
perhaps Merleau-Ponty's reflections on the work of psychiatrists Adhé-
mar Gelb and Kurt Goldstein—most famously the case of Schneider—in
order to articulate insights about ordinary perception and bodily self-
affection.[15] Unlike the phenomenological psychiatrists, Merleau-Ponty
is not primarily interested in offering a clinically useful analysis of—in
this case Schneider's—pathology. He is, rather, interested in the way
Schneider's experience reveals aspects of the taken-for-granted experience of normal subjects, specifically as this relates to embodiment. An
example here might be Schneider's paradoxical relationship with concrete and abstract actions. On the one hand, he copes better with concrete demands for action in his environment than with commands that
call for abstract actions or actions out of context. He can, for example,
react to an irritating fly by swatting it away more readily than he can
move his hand in response to his doctor's command. On the other hand,
his ability to follow commands is often accompanied by an inability to
respond to features of his environment that might otherwise be salient.
He recognizes his psychiatrist's (Goldstein's) house, but only when it is
his destination. If he passes it on some other errand, he simply fails to
notice it. In the first case, he seems excessively bound to the immediate
demands of his environment, while in the second case, he seems curiously
immune to that environment. On closer inspection, however, it turns out
that the two moments are consistent within the experience of Schneider
himself and simultaneously reveal significant aspects of ordinary embodied being-in-the-world. What Schneider lacks is the capacity to "reckon

with the possible."[16] That is to say that he lacks the capacity to see his environment as a field of *possible* action in which what is salient can change. In our ordinary comportment in the world, we perceive our environment as a space for the execution of plans and projects, but also as a space that continuously offers possibilities for acting in ways that were not conceived of in advance of our inherence in the situation. Schneider deals with his environment as demanding immediate action (swatting the fly) or enabling the execution of plans (going to Goldstein's house), but it fails to exert a significant influence on him as a dynamic space of affordances. The Schneider case reveals, among other things, that action is not primarily to be understood in terms of the unfolding of projects and plans upon an environmental substrate, but as a constant interface with a changing environment that both transforms and is transformed by intentional, embodied agency.[17]

So the pathology, as studied by phenomenologists, is doubly disclosive in the sense that it allows for a better understanding of the unique features and experiential presentation of this or that pathology (schizophrenia, depression, dementia, brain injury etc.), and because it reveals aspects of experience that are taken for granted in the ordinary being-in-the-world.

A further point that is important to stress here is that while the clinical and philosophical insights at stake are perhaps separable for the researcher, they are often one and the same in the experience of the sufferer, at least tacitly. As is clear from the descriptions of, for example, schizophrenia in the literature, the suffering is to a great extent constituted *by* the felt absence of the normal. Thomas Fuchs cites patients who speak about the distressing sense of a dissociation of a sense of ownership (SO) from a sense of agency (SA) in their own actions. If I am pushed from behind, my body moves. While it is obvious to me that it is *my* body that is moving and there is, therefore, a sense of ownership of the movement, there is no sense of agency, as I did not initiate the movement. This is just another way of saying that the movement cannot be considered to be my *action*. If I reach for my coffee, on the other hand, the experience is accompanied by both an SO and an SA. For many schizophrenics, on the other hand, SO and SA are dissociated even in cases where they *should* cohere. The sufferer reaches for the coffee, and thereby seems to be acting, but feels that the movement was initiated from elsewhere. In the terms in which we have been speaking, this simple example both describes a crucial moment of schizophrenic pathology and simultaneously clarifies a fundamental aspect of ordinary experience that often goes unnoticed. But it also discloses the co-belonging of these moments as constitutive of the actual suffering itself. That is, the schizophrenic does not just experience the dissociation of the SO from the SA, but experiences this as distressing. In other words, his/her suffering is constituted partly through a felt sense of what should constitute the action but does not.

We can conclude from this that the structure of normality is fundamentally constitutive for all lived experience, not just in terms of the regularities and continuities, so well described by Husserl in several studies,[18] but by a sense of how things *ought* to be. Things tend to be "thus and so," both in terms of the objects and horizons of perception themselves, and in terms of my comportment toward them. This establishes expectations not just of how things will be in the future, but in relations with self, others, and world ought to be in the present. Normal lived experience seeks what is optimal because it is optimal*izing* such that what deviates from the optimal is experienced as absent. This point is often framed within accounts of intersubjectivity and the distinction between cultural worlds, where the home world is encountered as the sphere of the normal, while the alien world presents as abnormal.[19] However, there is also a clear sense in many of Husserl's writings that normality can characterize in the more global sense I am referring to here.[20] Of course, deviation from the optimal is not in itself enough to constitute an experience of suffering. The deviation will have to reach a point where it disrupts an agent's capacity to act in the world to a significant degree for this to be the case. And when it does, the deviation protrudes in our experience in a way that forces us to attend to it. The normal *qua* normal is what *dis*-appears in ordinary world experience, but *dys*-appears, to use Drew Leder's term, in pathological experience.[21] Normality, as such, is not one pole of a normal/abnormal dichotomy—since this would imply a third position from which these both could be viewed—but is the ground upon which pathology shows itself and suffering makes sense.[22] The sufferer of schizophrenia not only experiences a dissociation of SO from SA, but experiences this *as abnormal*. In other words, while the description of pathology may allow us to see the normal structure of world experience, it is not necessary in order for us to experience it. To the contrary, the normal is the background, both conceptually and experientially, against which the abnormal is always profiled. The fact that schizophrenics feel disembodiment as suffering suggests that they retain an implicit sense of normal world experience which they experience as absent.[23] Their suffering is suffering because it involves insight into what is absent.

The schizophrenic experiences her thoughts as not her own, her body as not her own, or her actions as not her own. Such formulations bespeak suffering as a sense of the self as threatened or fragmented, as experiencing a split between what is and what should be. The "I" who experiences itself as broken apart and undermined by alien presence, experiences itself as a threatened unity and its suffering only makes sense in this way. In other words, the pair normal/abnormal is not simply a matter of categorizing experience from the outside but is immanent to experience as such, to the extent that the suffering of the schizophrenic is partly given through a felt sense of the absence of the normal in her world experience. The insight that pathology offers in light of phenomenology is complex, but is perhaps most importantly to be understood as the sense of suffering as such as it is lived by the sufferer.

The Phenomenology of Illness and the Idea of the Recoverable Normal

I want now to turn to another of the ways suffering and normality have been conceptualized within the phenomenological tradition. This second line of inquiry is, paradoxically, also largely Merleau-Pontyian in inspiration. This is paradoxical because the claims made here will, at least initially, appear to be in tension with the conclusions of part 1. In the first section, we claimed that (i) suffering is understood as a loss of the normal; (ii) that this loss discloses the normal as what is absent in the experience of suffering and (iii) that this disclosure is immanently constitutive of the experience for the sufferer. It is, in fact, only because of this third point that the relationship between suffering and normality can be articulated in the way it has been by phenomenological researchers.

The second approach is one developed by researchers into the phenomenology of illness, whose focus is not so much on what is constitutive of the meaning of suffering and normality, but on how these experiences are negotiated by subjects in the lifeworld. I have in mind, here, work done by Fredrik Svenaeus, Havi Carel, Kay Toombs[24] and others, and while I will not be able to treat the work of any of these in any depth for reasons of space, I hope to draw out certain fundamental points which are common to their respective treatments of illness and health as these relate to my topic. Rather than trying to contribute to clinical descriptions of this or that pathology and/or using these to delineate the normal structures of experience, these researchers have tended to focus on suffering as a way of being-in-the-world which affects all of us some of the time and, sadly, some of us most of the time. The point is not to describe suffering under the aspect of specific manifestations/conditions, but suffering as part of the lifeworld of existing subjects. That this second approach places suffering more globally in terms of being-in-the-world will also have implications for the structure of normality. Chief among these is the idea that normality is not simply banished in suffering, but can be reconstituted through modifications in the lived body's relationship with itself, with the world, and with others. This makes the notion of the normal a more dynamic and a more flexible one than some of the conclusions to the first section might have seemed to suggest.

I said that this second form of phenomenological attention to suffering is not concerned with articulating normal experience. This is not entirely true. Fredrik Svenaeus's discussion of illness and suffering, for example, is just as much concerned with elaborating the phenomenon of health as that of illness. Svenaeus describes health in terms of a homelike rhythm of being-in-the-world which is disrupted in illness.[25] Following Heidegger, he thinks of this homelikeness of health as an attunement to Being which is not a specific feeling, or even a mood, but a way of being-in-the-world, which is, in fact, our default way of being-in-the-world, a point that Svenaeus clarifies by contrasting the attunement of health with the mood of illness.[26] In other words, illness as disruption is better thought of as a mood that clouds a more fundamental attunement which itself only becomes visible when the disruptive mood of illness obtrudes. As our most fundamental attunement to Being, health hides in plain sight, as it were, which is precisely why a focus on the phenomenon of being ill is so fruitful in bringing it to light.

This means that Svenaeus' phenomenology of illness is doubly disclosive in a way that reminds of what we saw in the elaboration of pathology. Suffering provides, once again, a *via negativa* to normal world experience. In this case, the explicit focus is on health rather than normal experience, but the parallel is clear.

But there is more going on here which we did not see in the earlier discussion. Svenaeus and others working on the phenomenology of illness exhibit a clear interest in the implications of suffering and illness for living subjects, rather than simply what this suffering shows us. Normality obtrudes in its absence for the patient, but what does this mean? Is the story of illness a story of unqualified tragedy and of pure loss? Illness is, without doubt, "a disruption of the lived body"[27] in which the biological body obtrudes and disrupts the unity of the lived body, thus making ordinary ways of being difficult, if not impossible. But this experience of breakdown also enables a reconciliatory movement[28] in which the biological body is reintegrated in the unity of body and world such that what Svenaeus called the fundamental attunement to the world can be at least partially reconfigured. In making this claim, the phenomenologists of illness reveal a clear debt to Merleau-Ponty and specifically his emphasis on the plasticity of the living body and the capacity of the subject to cope with alterations to environment and body. There is also a debt to Kurt Goldstein's ground-breaking work *The Organism* from 1934,[29] which was, in many ways, as inspirational as Husserl's phenomenology for the evolution of Merleau-Ponty's thought on the subject of embodiment. In that work, Goldstein develops what might be termed an environmental approach to understanding illness and health and therewith also what can be meant by the term "normal." He says,

> An organism that actualizes its essential peculiarities, or . . . meets its adequate milieu and the tasks arising from it, is "normal." Since this realization occurs in a specific milieu in an ordered behavioural way, one may denote ordered behaviour under this condition as normal behaviour.[30]

There are two points worth noting here. The first is that the normal is not understood as a condition of the organism, but as a relation between the subject and its environment. Changes in either the subject or the environment can account for a loss of the normal, but it is the relation between them rather than either organism or environment independently that is paramount here. And what is more, because the normal is not a state of the organism, it can be restored or reconstituted without restoring the organism to its condition prior to loss or injury. Instead, restitution can come by means of modifications in the ways in which organism and environment interact.

The second point, which follows from the first, is that the normal is understood in terms of ordered behavior, which is to say that normality can be said to be present whenever the organism is able to comport itself in ways that are consistent, manageable, and coherent. Normality can be lost, but it can also be regained by the reorientation of the subject in her environment. According to Goldstein, then, becoming well again involves the establishment of a new individual norm[31] in which ordered behavior becomes possible again.[32] This can be a reversion to the original constellation of organism and environment, but it is more likely to be accounted for by ways of coping in which order is restored despite the changes in the organism or the environment.

The most important implication of Goldstein's claim is that the normal is not a fixed category, but a dynamic designator of the relationship between an organism and its environment. What is normal today, may not be so tomorrow. This means, in turn, that suffering should not be understood as the loss of the normal, but perhaps better as a disruption that requires a reorientation of the organism or a transformation of the environment in ways that will facilitate the return of ordered behavior.

There are, of course, a number of different ways in which such transformations or modifications can take place. One of the most interesting of these—and here we come closer to Merleau-Ponty's original contribution—is the body's plasticity and capacity to modify itself in order to cope with challenging environments. This is a capacity of human embodiment generally, but one that becomes especially conspicuous in cases of illness, injury, or other traumatic disruption. One might think, here, of the blind man's cane which transforms the inorganic matter of the cane to make it part of the sensing body and thereby circumvents the loss of vision.[33] The cane is the symbol of the loss of function, but is simultaneously the very manifestation of its restitution.

In a similar vein, Havi Carel gives the example of the wheelchair which, like the blind man's cane, at once both reveals the loss of a capacity for being, at the same time as it makes possible a reconstitution of normal being-in-the-world, through the incorporation of a foreign object into the body schema.[34] The wheelchair does not replace the ability to walk and nor does it negate the meaning of the loss that occurs in paralysis, any more than the blind man's cane replaces or trivializes what he has lost. What they do, rather, is witness to a capacity to reconstitute bodily unity and the relation between body and world in imaginatively novel ways. These reconstitutions are the new norms of which Goldstein spoke. The paralyzed legs are a lacuna in the normal—objectified and obtruding as they are in the context of the wholeness of the body schema—but at the same time, mobility is restored through the incorporation of the foreign object, the wheelchair. And by the same token, the wheelchair, once inanimate, becomes an integrated part of the body's internal unity and of the lived unity of body and world. Suffering bespeaks loss and disruption, but it also makes possible an imaginative adaptation in which the normality of experience can be reconstituted even if it cannot be simply recovered.

In addition to the phenomenological accounts we have been discussing, such reconstitution is also testified to in a number of clinical studies. Neurophysiologist Jonathan Cole, who worked extensively with spinal injury patients tells us that "many (disabled) people . . . have explored what is possible, bringing their remaining capacities and capabilities to make a whole, their whole,"[35] while Alexander Luria describes a brain-damaged patient who "continues to try to recover what was irretrievable, to make something comprehensible out of all the bits and pieces that remain of his life."[36] These and other examples, reveal the fragility of human subjects, but also their capacity for adaptation and the reconstitution of the normal.[37] The normal is not a condition that is present or absent, but denotes a relation of coping which obtains in the interface between the subject, the world and others.

These examples have, therefore, important implications for the way we understand the structure of normality in experience and its relation to the meaning of suffering. In the first part of the chapter, we saw that normality was both revealed *by* pathological experience as well as being revealed *through* such experience in a way that often accounted for the meaning of the suffering of mental disorders. The schizophrenic, for example, is conscious of the absence of a normal sense of agency and experiences this as distressing. On this account, though, the suffering is, we might say, unredeemed. Suffering is the present absence of the normal, while the normal is understood, in a sense, as the absence of suffering. But the phenomenologists of illness have drawn a more complex picture, in which the loss of the normal in suffering breaks apart but also simultaneously makes possible a reconstituted unity of subject and world. Taking a cue for Merleau-Ponty's notion of the plasticity and dynamism of the body schema, this approach understands suffering very definitely as loss, but also as figurative of embodied being-in-the-world's capacity to adapt and reconstitute the normal. In other words, while the first approach tends to present normality as a more or less static aspect of experience, whose absence designated the meaning of suffering, the second approach identifies a dynamism in the subject and also consequently in the notion of the normal.

Toward a Resolution: The Embodied Constraints on the Recoverable Normal

The reader may be left with the impression that what we have here is two alternative accounts of the meaning of suffering viz the structure of normality in experience. In fact, this is not the case. What we have seen instead is simply different points of focus. The psychopathologists are interested in the structure of the experience of specific phenomena as well as what these reveal about normal world experience as such. Their concern is not the individual patient and his/her capacity to cope with his/her suffering. But precisely our capacity to cope with suffering is the point for researchers in the phenomenology of illness. Now, one might perhaps be tempted to say that precisely the lack of (research) concern for the patient's response to suffering has led to a blindspot in the work on psychopathology inasmuch as the notions of normal and abnormal tend to be conceived of as mutually exclusive. However, I would suggest that this critique only really makes sense if we overestimate the conclusions of the second section.

What we saw in the second section was that the normal is not a state of the organism, but a complex and dynamic relation between the organism and its world. Furthermore, we showed the extent to which this relation can be reconstituted in spite of loss, injury or disruption. But having said this, we must be careful not to overestimate the capacity for reconstitution or the implications of reconstitution for suffering. Goldstein himself makes this point in relation to the capacity of animals and humans to adapt to their environment following injury. He says that the following:

> It is self-understood that animals, after amputation of limbs, cannot cope with all the demands that they "normally" could meet. One easily overlooks these limitations, because one pays attention . . . to the restoration of particularly important performances. For example, one pays attention to the restoration of locomotion in animals after amputation . . . or of the function of a certain muscle after transplantation in human beings, and so on. We know that, after transplantation, the restored energy is seldom greater than one third of the energy of the control muscle and that the transplanted muscles suffer abnormal fatigue in the originally "normal" performances.[38]

In other words, our fascination with the restorative capacities of the organism can easily blind us to the loss of the optimal way of being-in-the-world that is not recovered. The organism can adapt itself and/or its environment in order to restore the possibility of ordered behavior, but even this never entirely overcomes what is lost and suffered. To the contrary, reconstitution can involve an increase of suffering for the sake of restoring function, as when the blind man choses the added difficulty of making his own way in the world rather than choosing the easier path of giving in to the limitations imposed by his lack of sight. Reconstitution is, in other words, purchased at the price of additional suffering.[39] So the normal is reconstitutable, but not endlessly so, because organisms live within forms of organization that are constituted within frameworks of optimal and sub-optimal. Georges Canguilhem tells us that the notion of normal is a conceptualization of human being-in-the-world and while this is no doubt correct, it can mislead us into thinking that the conceptualization is as endlessly malleable as language itself.[40] That this is not the case is evident from both the phenomenological analyses and the clinical studies to which we referred in part 2. All spoke of extraordinary capacities of recovery, but never in ways that could be read as simply obviating the original loss of the normal. Cole's recovered wholes, Merleau-Ponty's cane and Carel's wheelchair are all examples in which the normal is both recovered *and* experienced as absent at the same time. Suffering is not abject because it is not pure, unmitigated loss. But nor is suffering entirely mitigated by the adaptive capacities of the sufferer. This is very clear in Carel's insistence that Merleau-Pontyian plasticity amounts to ways of "living with" suffering, not of neutralizing it.[41]

That phenomenologists of illness speak of recovering the lived unity of being-in-the-world does not mean that suffering is overcome or that normality is endlessly configurable. It means only that the loss of the normal can be attenuated. In the same way that the plasticity of a text is not infinite but bounded by both its own inner coherence as well as by the capacity of the reader, being-in-the-world is bounded by the structure of normality that is itself indexed to the meaning-constituting structures of subjectivity. This structure can be disrupted by illness, pathology etc., such that the disruption is experienced as suffering. As the psychiatrists show us, this suffering can offer insight, into the suffering itself and that from which the suffering is a departure. And as the phenomenologists of illness argue, the loss of the normal in suffering is not a monolith, but can, in a sense, be redeemed through the subject's capacity to reconfigure its relation to its environment and thereby regain a new normal. But such reconfiguration can never entirely escape the logic of the optimal relationality of organism and world. At root, suffering is disclosive and what it discloses is an obtrusive lack at the heart of our present being-in-the-world that mourns the loss of what it lacks.

On the basis of this, we may be confident in making at least the following three points:

(1) That normality is an ineradicable regulative framework for experience and meaning. Even to the extent that the normal can be

reconstituted, it remains the orientation of all organisms to seek the kind of ordered coherence with their environment that we have described under the designation "normal."

(2) That while the normal can be reconstituted, it is not an endlessly malleable designation because it is bound to forms of organization— social *and* material—that are not just social negotiations but indexes of embodied subjectivity.

(3) That since reconstitution is only ever partial, suffering and normality are not mutually exclusive in the sense that normality is often recovered in ways that amplify suffering and the felt loss of optimal being-in-the-world.

To conclude, we have seen several ways in which phenomenology contributes to our understanding of suffering. On the one hand, suffering as loss is mobilized in order to help us to see what is experientially closest to us, but yet hardest to say. On the other hand, the coping body subject is seen as possessing the capacity to reclaim a sense of wholeness in the face of the loss in such a way that neither the lived unity of being-in-the-world, nor suffering as such are negated but are preserved together. In sum, the phenomenological contribution is significant because of its insistence on the co-belonging of subjective and objective moments in understanding suffering and the constitution of normality. It presents us with a challenging and complex account of suffering itself as well as its meaning for self- and world-relation. But if this account of the complexity of the relationship between suffering and normality is at all warranted, which I have argued it is, then the phenomenological account also offers an important corrective to any one-sided attempts to address suffering only through word (constructivism) or deed (biomedical paradigm).

Notes

1. I say "aspects" of world experience to avoid the impression that I mean that ordinary world experience is free of suffering. The point is rather that specific kinds of suffering can make us aware of dimensions of experience of which we tend otherwise to be unaware.
2. Karl Jaspers, *General Psychopathology*, 2 vol. (Baltimore: Johns Hopkins, 1997).
3. Ibid., 2.
4. Ibid., 12.
5. See, for example, American Psychiatric Association's, *Diagnostic and Statistical Manual of Mental Disorders* (5th ed.) DSM V. (Washington, DC: American Psychiatric Publishing, 2013).
6. Jaspers, *General Psychopathology*, 25.
7. This is similar to the arguments against reductionism put forward by analytical philosophers of mind such as Thomas Nagel and Joseph Levine, *Mind & Cosmos: Why the Materialist Neo-Darwinian Conception of Nature Is Almost Certainly False* (Oxford: Oxford University Press, 2012), 31; Joseph Levine, "Materialism and Qualia: The Explanatory Gap," *Pacific*

Philosophical Quarterly 64 (1983); Thomas Nagel, "What Is It Like to Be a Bat?" *The Philosophical Review* 83, no. 4 (1974).

8. Jaspers, *General Psychopathology*, 17.
9. Thomas Fuchs, "Corporealized and Disembodied Minds: A Phenomenological View of the Body in Melancholia and Schizophrenia," *Philosophy, Psychiatry & Psychology* 12, no. 2: 95–107.
10. Ibid., 103. See also Shaun Gallagher, *Phenomenology* (Basingstoke: Palgrave Macmillan, 2012), 131.
11. Louis Sass, *Madness and Modernism* (New York: Basic Books, 1992), 8. See also Louis Sass, "Self and World in Schizophrenia: Three Classic Approaches," *Philosophy, Psychiatry & Psychology* 8, no. 4 (2001): 266.
12. Sass, "Self and World in Schizophrenia: Three Classic Approaches," 262.
13. Dan Zahavi and Josef Parnas, "The Link: Philosophy—Psychopathology—Phenomenology," in *Exploring the Self: Philosophical and Psychopathological Perspectives on Self-Experience*, ed. Dan Zahavi (Philadelphia: Johns Benjamins Publishing, 2000), 9.
14. Martin Heidegger, *Being and Time* (Oxford: Blackwell, 1962), 104.
15. Maurice Merleau-Ponty, *Phenomenology of Perception* (London: Routledge, 2012), 105f.
16. Ibid., 112. See also Komarin Romdenh-Romluc, "Merleau-Ponty and the Power to Reckon with the Possible," in *Reading Merleau-Ponty*, ed. Thomas Baldwin (Abingdon and Oxon: Routledge, 2007), 44–58.
17. For an example of the kind of account that explains the difference between human and non-human action purely in terms of intention, see Elisabeth Anscombe, *Intention* (Cambridge, MA: Harvard University Press, 2000).
18. Edmund Husserl, *Experience and Judgement: Investigations in a Genealogy of Logic*, ed. Ludwig Landgrebe (Evanston, IL: Northwestern University Press, 1973), 122; Edmund Husserl, *Analyses Concerning Passive and Active Synthesis*, ed. Anthony Steinbock (Dordrecht: Kluwer Academic Press, 2001), 136–7.
19. Anthony Steinbock, *Home and Beyond: Generative Phenomenology after Husserl* (Evanston, IL: Northwestern University Press, 1995).
20. Ibid., 138–47. See also Cristian Ciocan, "Husserl's Phenomenology of Animality and the Paradoxes of Normality," *Human Studies* 40 (2017), 175–90.
21. Drew Leder, *The Absent Body* (Chicago: Chicago University Press, 1990).
22. Ciocan, "Husserl's Phenomenology of Animality and the Paradoxes of Normality," 176.
23. This is not just a feature of schizophrenia, but also holds for other disorders. For a discussion of a similar structure in the experience of dementia, see Michela Summa, "The Disoriented Self. Layers and Dynamics of Self-Experience in Dementia and Schizophrenia," *Phenomenology and the Cognitive Sciences* 13, no. 3 (2014): 477–96.
24. Fredrik Svenaeus, *The Hermeneutics of Medicine and the Phenomenology of Health: Steps Towards a Philosophy of Medical Practice.* (Dordrecht: Kluwer Academic Publishers, 2000); Havi Carel, *Illness: The Cry of the Flesh* (Stocksfield: Acumen Publishing, 2008); S. Kay Toombs, *The Meaning of Illness: A Phenomenological Account of the Different Perspectives of Physician and Patient* (Dordrecht: Kluwer Academic Publishers, 1993).
25. Svenaeus, *The Hermeneutics of Medicine and the Phenomenology of Health: Steps Towards a Philosophy of Medical Practice*, 94.
26. Ibid., 95.
27. Toombs, *The Meaning of Illness: A Phenomenological Account of the Different Perspectives of Physician and Patient*, xvi.
28. Carel, *Illness: The Cry of the Flesh*, 87.

29. Kurt Goldstein, *The Organism* (New York: Zone Books, 1995).
30. Ibid., 325.
31. Ibid., 333.
32. Ibid., 336.
33. Merleau-Ponty, *Phenomenology of Perception*, 144.
34. Carel, *Illness: The Cry of the Flesh*, 86.
35. Jonathan Cole, *Still Lives: Narratives of Spinal Cord Injury* (Cambridge, MA: The MIT Press, 2004), 277.
36. A.R. Luria, *The Man with a Shattered World: A History of a Brain Wound* (Harmondsworth: Penguin Books, 1975), 159.
37. See, for example, Oliver Sacks's, *The Man Who Mistook his Wife for a Hat and Other Clinical Tales* (New York: Harper Perennial, 1987); Arthur Frank, *The Wounded Storyteller: Body, Illness, and Ethics* (Chicago: Chicago University Press, 2013).
38. Goldstein, *The Organism*, 337.
39. Ibid., 341.
40. For a more linguistic approach to the constitution of the normal in experience, see Georges Canguilhem, *The Normal and the Pathological* (New York: Zone Books, 2015), 228, and also Michel Foucault, "Introduction," in *The Normal and the Pathological* (New York: Zone Books, 2015), 19.
41. Carel, *Illness: The Cry of the Flesh*, 88. This makes the phenomenological approach as different from social constructivism as from the biomedical paradigm.

10 Re-possibilizing the World
Recovery from Serious Illness, Injury or Impairment

Drew Leder

Prelude

Sometimes it helps to start with a story. As recounted by Jack Riemer in the *Houston Chronicle*, famed violinist, Itzhak Perlman, was launching into his New York concert when something went wrong. One of his violin strings snapped. Perlman walked slowly on crutches, a legacy of his childhood polio—would he now need to leave the stage in search of a new violin or string? Riemer writes:

> But he didn't. Instead, he waited a moment, closed his eyes and then signaled the conductor to begin again. The orchestra began, and he played from where he had left off. And he played with such passion and such power and such purity as they had never heard before.
>
> Of course, anyone knows that it is impossible to play a symphonic work with just three strings. I know that, and you know that, but that night, Itzhak Perlman refused to know that. You could see him modulating, changing, recomposing the piece in his head. At one point, it sounded like he was de-tuning the strings to get new sounds from them that they had never made before . . . and then he said—not boastfully, but in a quiet, pensive, reverent tone—"You know, sometimes it is the artist's task to find out how much music you can still make with what you have left."[1]

It should be noted that there are questions about the accuracy of Riemer's account. It was written down several years after said concert, and contemporary records show no reviewers referring to this onstage incident.[2] The story may be embellished or apocryphal. Nonetheless, we can use it as a powerful parable, one that will help us explore recovery from illness, injury and impairment. When a string snaps—when we lose bodily capabilities that before were present, thereby threatening our social, psychological and existential integrity—how can we still make music with what we have left?

I will take as my paradigm case of "string-snapping" the onset of a serious physical illness or injury that brings with it long-term restrictions and anxieties. There are a great number of pathographies, ethnographies and phenomenologies of bodily illness and impairment.[3] I will confine myself to a few recorded by unusually reflective observers. Without claiming to arrive at transcultural essences, I do seek to explore a few fundamentals that often characterize such experiences of disruption and loss. My focus, using the Perlman story as a clue, is how one *re-possibilizes* one's life in the face of newfound impossibilities.

While focusing on serious illness and injury, this analysis may prove applicable to a range of other predicaments; for example, living with a disability, or simply growing old. Ram Dass, the spiritual teacher who did much to bring Eastern wisdom to the West, suffered a late-life stroke, placing him in a wheelchair and damaging his language abilities. To cope well with his newfound limitations, and all the other challenges of aging, he learns from other people living with disabilities, and his own work with the dying. He writes, "Whether their bodies have been incapacitated by physical trauma, disease, or old age makes little difference, since the results are the same."[4]

Of course, to simply equate these situations can also cover over important differences. Some disabilities, such as deafness or blindness, may be present from birth. As those in the burgeoning field of Disabilities Studies have pointed out, these and other conditions that shift one's embodiment away from the mainstream, should not be equated with diseases. Much of the "dysfunctionality" associated with such conditions may be created by the larger society which misunderstands or stigmatizes these "non-ideal bodies," and constructs obstacles to their full social engagement.[5]

Just as having a socially defined disability should not be equated with disease, neither should being old. In the contemporary West, we live in a culture of pernicious and pervasive ageism, where elderhood is falsely equated with dotage, disability and diminishment. With the dramatic expansion of the human lifespan, many elders are leading healthy and vigorous lives, not to mention enjoying some of the fruits of wisdom and liberty that can grace the later years.[6]

Still, much said here may be applicable to those with certain disabilities, and/or facing old age. While recognizing that many defined as "disabled" may be nonetheless quite healthy, Wendell is concerned that we do not underestimate "the proportion of people with disabilities who are either disabled by what we would all recognize as illness, or ill as a consequence of disability."[7] She notes, for example, the large numbers of North Americans disabled by arthritis, heart or respiratory disease and diabetes. It is also true that old age, while not itself a disease, often brings with it various pains, impairments and chronic illnesses. These simply tend to accumulate as our cells and organs age.

In short, many life-conditions are biologically painful, dangerous or limiting, not simply the result of social discrimination against the "old" or "disabled"—though this discrimination can make things worse. At some point in life, perhaps most of us face the issue of "how much music you can still make with what you have left."

In harmony with the Perlman example, I give a somewhat musical organization to the chapter. The prelude now done, we turn to "three movements," followed by a coda in place of the usual scholarly conclusion. At times a composer will explore a coherent theme through the different movements of an orchestral work. I will do the same, speaking of illness and recovery as incorporating three moments. As reflected in the words chosen, these include notes of both likeness and opposition with each other: (i) bodily *"impossibility,"* (the loss of previous capabilities and identities); (ii) *"I'm-possibility" (reclaiming a liberty of creative response)*; and (iii) the *"'I am' possibility"* (the experience of an *I am*, a level of being that transcends the bodily and social limitations that illness/injury has made so apparent).

First Movement: *Impossibility*

"Of course, anyone knows that it is *impossible* to play a symphonic work with just three strings" (emphasis mine). "Impossible" is derived from the Latin *in* meaning "not," and *posse*, to "be able." Serious illness and injury can make one unable to do—or even be—things one previously was. Paul Kalanithi, in the midst of a highly successful life as a top Stanford neurosurgeon, describes how he suddenly finds himself incapacitated by metastatic lung cancer. In *When Breath Becomes Air*, he writes:

> My body was frail and weak—the person who could run half marathons was a distant memory—and that, too, shapes your identity. Racking back pain can mold an identity; fatigue and nausea can, as well. . . . Because I wasn't working, I didn't feel like myself, a neurosurgeon, a scientist—a young man, relatively speaking, with a bright future spread before him. Debilitated, at home, I feared I wasn't much of a husband for Lucy. I had passed from the subject to the direct object of every sentence of my life. In fourteenth-century philosophy, the word patient simply meant "the object of an action," and I felt like one.[8]

The impossibilities faced by an ill person are particular and personal in nature. Kalanithi is unable to run a half-marathon. This is not something illness, for example, would take from *me* because I have never contemplated, or even been fit enough, to run such a race. But Kalanithi was, and did, and loved it. He was an accomplished neurosurgeon at work, an engaged husband at home—yet these personal capacities now too lie in rubble.

Thus the impossibilities brought about by serious illness/injury are not *logical* in form—that "A" cannot equal "not A"—nor merely *imaginative*—that Kalanithi cannot sprout wings and fly—nor the impossibilities that might *belong to another*—for example, the inability to do long-distance trucking does not worry Kalanithi (never a truck-driver), though it might devastate someone who lived by that trade. No, illness-impossibilities are *specific to the individual*, based on the activities they have engaged in or foreseen, which now are blocked. For Perlman the question is can he play his concerto with just three strings; for Kalanithi, whether he can be a runner, lover and surgeon in the face of metastatic lung cancer.

But, as this account suggests, serious illness threatens more than a repertoire of specific actions—it can twist one's very identity. Kalanithi no longer feels the "subject" at the center of a radiating life, but rather the "object," the passive "patient," to whom things are done. Kevin Aho, in this volume, talks of his massive unexpected heart attack at age 48 as "nothing less than world shattering. The visceral awareness of the frailty and vulnerability of my body and the recognition of my fundamental dependency on others have trashed the illusions of my own strength and autonomy."[9] Serious illness can disrupt one's relationship with one's own body, now an antagonist. While experiencing an increased need of others, one may also feel excluded, exiled to the island of the sick or dying.[10] One's sense of cosmic meaning may be upended (Why would God do this, is there even a God?), as is the narrative of one's unfolding life. Aho writes, "When the future collapses in this way, the resources of the past that I relied on to create my identity no longer held, and I was left to confront the ultimate question, '*Who am I*'? I was no longer the productive professor, the cyclist, or the skier."[11]

The difficulties associated with newfound impossibilities—of doing what one did, or being who one was—are vectored by the response of others, and not always in a good way. In the midst of a vigorous midlife, Susan Wendell developed a form of chronic fatigue immune dysfunction syndrome (myalgic encepalomyelitis). Because this disease is largely invisible to others, she found herself trapped in a two-sided dilemma; either seek to power her way through the disability "like a normal person," without receiving understanding and accommodations from others—or draw attention to her condition and feel like a complainer in search of sympathy.[12] There is thus a price to be paid for having *invisible impossibilities*.

But so, too, is there for those that are highly *visible*. Nancy Sherman, writing of soldiers who have lost limbs in combat, notes that "amputees often feel stigmatized by their injuries, self-conscious that others view them as freakish or disfigured or just different."[13] She recounts the story of Dawn Halfager who had her right arm blown off in an explosion in Iraq.

> She told herself, "I'm never going to be normal. It's never going to be easy. . . . When I lost an arm, that went out the window." There is a tension in her images of herself, then and now, of who she would like to be and who she is, filtered through the eyes of others and her perception of their gaze, and through her own grief.[14]

Though Dawn makes dramatic progress, is the CEO of a large company, and an avid tennis player, she cannot have the body or be the person from before, as the gaze of others ceaselessly reminds her.

Having a body with "only three strings," using the metaphor from the Perlman story, is also made more difficult by our social idealization of the "perfect body." As Wendell writes,

> Body ideals include not only ideals of appearance, which are particularly influential for women . . . but also ideals of strength, energy, movement, function, and proper control; the latter are unnoticed assumptions for most people who can meet them, but they leap to the foreground for those who are sick or disabled.[15]

Incapacities come to feel like shameful inadequacies in the face of the pressure to fulfill these tight norms.

This may be reinforced by an experience of social shunning whether of the person, or the topic of their disease. Havi Carel one day found herself short of breath: a workup revealed the devastating news that she had lymphangioleiomyomatosis (LAM), a severe, progressive and, at the time, uniformly fatal disease. She found this burden was increased by other's discomfort and denial. "With many the fact of my illness is never mentioned and I have come to experience it as something highly secretive, grossly inappropriate for conversation. My dirty little secret."[16] This is echoed in Kevin Aho's account:

> What was especially disturbing is that some colleagues, whom I considered friends before the illness, appeared to avoid me altogether or would simply smile and scurry away, perhaps uncomfortable with what I represented, shattered health, vulnerability, and a reminder of death.[17]

One would hope that at least medical professionals would surmount this aversion. But encountering bodily *impossibilities* can be as hard for them as well: they may be disappointed in the ineffectiveness of their treatments, fearful of failure, or of death itself, and even turn the blame upon the patient. Arthur Frank, the well-known sociologist, recovering from both a heart attack and testicular cancer, writes: "In society's view of disease, when the body goes out of control, the patient is treated as if he has lost control. Being sick thus carries more than a hint of moral failure; I felt that in being ill I was being vaguely irresponsible."[18]

Movement Two: *I'm-Possibility*

Though starting with the theme of *impossibility*, that is hardly where the story ends. Again, Riemer wrote:

> Of course, anyone knows that it is impossible to play a symphonic work with just three strings. I know that, and you know that, but that night, Itzhak Perlman refused to know that. You could see him modulating, changing, recomposing the piece in his head. At one point, it sounded like he was de-tuning the strings to get new sounds from them that they had never made before.

In the face of obstructions from illness and injury, the recovery process involves a reclaiming of a certain existential stance—that of *possibility*. Classic existentialist philosophy, for example, that of Sartre, emphasizes strongly the freedom that distinguishes the human being as "for-itself," from an "in-itself" material object governed by deterministic laws.[19] Though humans have a factical life situation, they always transcend it: they retain the capacity to define the meanings and purposes of their life, and choose the actions they will pursue. In this sense, even bodily impossibilities are embraced within a larger framework of "*I'm-possibility*"—an identity constituted of freedom precisely from any pre-determined identity.

However, this dualism of the for-itself and in-itself, of transcendent consciousness and the materiality to which it is conjoined, fails to acknowledge sufficiently the truth that illness reminds us of; that we *are* our bodies, our freedom arising directly from our embodied capabilities. As Sartre's colleague, Maurice Merleau-Ponty, explored in depth, there is a generalized bodily "I can," a structure of ability which enables us to experience and act in the world, and which manifests in multiple, specific capacities.[20] At the moment I am sipping on a cup of coffee I brewed earlier in the kitchen; caffeinated energy flows through me, helping me type, delete, rewrite before I finally finish, and take my dog for a walk on this lovely spring day, no doubt greeting neighbors along the path. All of this activity—gustatory, energetic, intellectual, sensorimotor, social, linguistic—is based on the specific capacities of my mouth, larynx, fingers, neurons, eyes, legs—not only qua anatomic structures, but as educated into certain habitual performance-capabilities.

So what does one do when "a string breaks," and the concerto that is one's life becomes disrupted? Often the ill or injured individual, after a keen experience of new *impossibilities*, sooner or later begins that process of re-habituating the body: "modulating, changing, recomposing the piece . . . de-tuning the strings to get new sounds from them." This can take many different forms.

First, one might learn to use one's injured body in a new way, accomplishing one's goals through different sensorimotor pathways. Finding oneself unable to climb into bed due to a weakened or paralyzed leg, one might learn to reach down and lift that leg up with one's arms. As sight worsens one may draw more keenly on a sense of hearing and touch when moving through a room. Carel, adjusting to her severe respiratory disease, writes: "I adapted to my breathlessness by finding circuitous routes to the shop to avoid walking uphill; I pause several times while climbing up a flight of stairs. . . . Finding a new way of performing an old task, given an altered set of capacities, is challenging; successful performance leads to a sense of achievement."[21] Like Perlman, Carel finds a way to complete the recital though playing on the "three-stringed violin" of her altered body.

I have been comparing Perlman's violin to a lived body (later, I will say more about Perlman's own physical impairment). But this instrument when viewed *simply as violin* has its own teachings. Without it, Perlman would not have his musical expressivity. Through his long years of practice, the violin has become, as it were, part of his lived embodiment, not simply an outer object, but the expressive voice through which his fingers skillfully sing.

In a famous passage, Merleau-Ponty writes of how a blind man experiences his stick not as an external object he touches, but that *from* which he touches, sees, and negotiates the world.[22] The stick has been *incorporated*—etymologically "brought within a body"—into the blind man's body schema, now part of the sensorimotor structure of his "I can."

Thus one also re-possibilizes the ill and injured body—through supplemental tools and technologies. An individual rendered paraplegic by a car accident can be re-possibilized by a motorized wheelchair, enabled to resume crucial life activities. A woman who is growing deaf may find assistance in a hearing aid, or even the higher-tech cochlear implant.

Yet before valorizing the isolated cyborg lived body, we might ask where this wheelchair, this cochlear implant, came from. They only manifested because of a perceived social need, and an investment of social resources, that made it available to the impaired individual. Nor should we forget professionals and caregivers who help train one in the use of a technological prosthetic, or those who build appropriately designed environments and access corridors. This reminds us that, whether it involves technology or not, re-possibilizing the ill and disabled body is a *social task*. To the extent we understand that we are all but temporarily abled—if we live long enough it is likely that we will face a variety of impairments—we can see it is in all our interest to provide social support and resources for the disabled body.

We find this implicit in our example: Perlman needs a symphony orchestra to play with him—and to wait patiently while he retunes his violin. So, too, is the appreciation of the audience needed, and a concert hall that brings them all together and is acoustically built to let the music sing. Similarly someone with a serious illness, injury or impairment needs medical and financial support; competent and caring professionals; understanding friends and family; workplace policies that incorporate medical leave and flexible work-hours; architecture, elevators, ramps, etc. that assure accessibility and so on. Earlier, I discussed how social rejection can intensify the experience of illness-as-*impossibility*; conversely, social support can enhance the "*I'm-possibility*" experience, showing there is, or should be, a "we" involved.

Nonetheless I retain the term "*I'm-possibility*" for a specific reason. In some ways, this "second movement" of this piece seems the reversal of the first: the rediscovery of personal *possibility* relieves the horror of the *impossibilities* brought about by illness. But the one mode doesn't eradicate the other. Despite the difference in punctuation marks—an apostrophe, a hyphen—*I'm-possibility* in fact remains the same word as *impossibility*. This is to register that, unless immediately and fully repaired, a serious illness or injury generally leaves its mark. The recovery process usually does not fully undo rupture, leaving us unchanged. The orchestral piece that Perlman played with his three-stringed violin is not the same as that he would have played before his string broke. It is unlikely that he was able to produce the identical notes, but even if so, these would not have had the identical meaning: they were now inflected by the marks of disruption, and all that Perlman needed to draw on—experience, persistence, flexibility, creativity, even courage and humor—in order to carry on. "You know, sometimes it is the artist's task to find out how much music you can still make with what you have left." The artist asserts "*I'm possibility*," but its meaning doesn't only surmount, but includes, the experience of *impossibility*, the element of the broken string.

This is the "artist's task" for those who face serious illness or injury. Recovery involves struggle and alteration, but can also manifest as the *acceptance* of ongoing limitations. As Ram Dass says, if you have turned into someone with arthritis, one often simply does best to "give up the model of being somebody without arthritis."[23] It is what it is. The fundamental condition—in this case, arthritis, which can be painful and limiting—persists. But there are associated modes of suffering compounded of fear, frustration, self-pity, and resistance, which can now drop away. With acceptance comes a peace which lessens emotional pain, and perhaps even the physical. This too is a way that with "three strings" one yet continues to play and hear life's music.

Along with accepting diminishments, one may discover that the illness has brought certain gifts, just as Perlman's performance was portrayed as enriched by his broken string. Carel describes how her severe and presumably fatal illness activated a new intensity of life:

> I no longer save money. . . . I care much less what people think of me. . . . I found to my surprise that I experienced amplified enthusiasm and joy, echoing against the narrow confines of the present. All my energy and happiness are funneled into the *now*, into today: how nice it feels to be *here*, in the sun, having a massage, listening to beautiful music, laughing until I am dizzy, sitting by a warm fire, experiencing friendship, love, sunshine, the lazy sensation of waking up after a deep sleep, the sharp authority of beauty.[24]

Sometimes this life-enrichment can even look the opposite of recovery as medically defined. After his stroke, Ram Dass is pushed by well-meaning therapists to regain his speech and motoric capacities. He writes:

> All the therapies call upon me as an Ego: Try harder! Don't you want to get better? Exert your will! . . . The stroke became a playing field for a whole new level of achieving: How much "progress" has there been? Can you walk yet? More gold stars to be won. Instead of will, I've found in myself a peaceful surrender to the karmic unfolding of my life—an unfolding that's like a tree growing or a flower blooming.[25]

Long a famous and frequent public speaker, Ram Dass learned to enjoy silence, and the essence that emerged when verbiage was stripped away. Long an itinerant wanderer, he found that the stroke grounded him, bringing him home to his own body—even to an appreciation of simply sitting in a wheelchair rather than struggling to regain his ability to walk.[26] It would be a limited reading of the Perlman story to view it as the tale of a triumphant act of the lone hero who, as Riemer says, "refused to know" limitation. Sometimes we triumph more through acceptance than refusal.

This also involves accepting the help of others. Being in the dependent position, though a blow to the ego, can allow us to grow more vulnerable and intimate. As Aho writes:

> The people in my life that I ordinarily took for granted became luminous and fragile. And this extended beyond my family, partner, and intimate friends, but to colleagues, neighbors, and even complete strangers at the supermarket or gas station. The masks came off, and I sensed how helpless and dependent we all are, and this made me love even more.[27]

For Aho, the illness serves as a teacher not simply of the head, but of the heart, as he feels in an embodied way his connection with, and compassion for, others.

Movement Three: The *"I am"* Possibility

The last movement of a symphonic work may spin things in a new direction, perhaps employing a novel rhythmic structure or musical form. At the same time it may refer back to themes earlier introduced. This last, brief movement will play with both summary and difference.

I have stressed the embodied nature of changes induced by illness/injury. The broken body brings painful notes of *impossibility* into a habitual life. Yet the person is also capable of reclaiming the stance of *"I'm-possibility,"* not in abstract form, but as an active re-tuning of their "three-stringed" body.

But our relation to this body can be peculiar, paradoxical.[28] Just when sickness shows us how inescapably we are embodied, this body also surfaces as something alien, aversive, hard to understand and control. It is that which limits our agency, may threaten our very existence. This body, at once too close and too distant, pushes us to seek what *transcends* its limitations. For some, this involves religious meanings; others may expand horizons by intellectual work, feeling the embrace of a loving community, or losing ourselves in an engrossing work of fiction. In Carel's words, "Even in cases of extreme physical disability there is always a freedom of thought, imagination, emotion and intellect."[29]

Recognizing such modes of transcendence need not imply a metaphysics of mind-matter dualism, or a rejection of the body. As Wendell writes, drawing on her life with chronic myalgic encaphitis,

> To choose to exercise some habits of mind that distance oneself from chronic, often meaningless physical suffering increases freedom, because it expands the possibilities of experience beyond the miseries and limitations of the body. . . . Nor do I think we need to devalue the body or bodily experience to value the ability to gain some emotional and cognitive distance from them . . . since it is bodily changes and conditions that lead us to discover these strategies. . . . Thus, the body itself takes us into and then beyond its suffering and limitations.[30]

Ram Dass, a long-time spiritual explorer, applies a different interpretive framework to his own experience of transcendence:

> What was changed through the stroke was my attachment to the Ego. The stroke was unbearable to the Ego, and so it pushed me into the Soul level also, because when you "bear the unbearable," something within you dies. My identity flipped over and I said, "So that's who I am—I'm a Soul!" . . . There's a paradox here, because although I'm more in the spirit now, I'm also more human. . . . Now I can risk going deeper into my incarnation, because I'm feeling more secure in my identity with the Soul.[31]

Ram Dass does not reject his body; his sense of security allows him to embrace his incarnation. But he also experiences himself as not *just* a body, such that he no longer fears old age, incapacity and death. He writes, "my Awareness doesn't have a locus, and so my consciousness isn't trapped in my body. I had *experienced* that—not as an abstract understanding, but as a real event."[32] Ram Dass was truly a three-stringed violin, his speech and motoric capacities profoundly compromised by stroke. Yet this enabled a new music.

Whether this exploration of transcendence is understood as fundamentally embodied, disembodied or somehow participating in *both*, I will refer to this as the "'*I am*' possibility" that illness may provoke. In the tradition of *jnana yoga* (the Hindu path of wisdom), one proceeds by progressively detaching from a limited sense that "I am this," or "I am that." Whatever predicates defined one's identity—I am tall, I am hungry, I am a good student, I am shy—could be altered, and probably has been during the course of one's life—yet a sense of present awareness remains, the primordial sense of "I am-ness." Here, a *jnana yogi*, would say we contact transcendent Being. Illness, injury, incapacity and old age clarify what is passing in our personal identity, and yet what transcends this and remains.

A counter-example to this "I am" experience might seem to be found in the Buddhist teaching that nothing in the universe has substantial and separate reality: the I, the ego-self, never really exists, and belief in this fiction is the source of our sufferings. But this tradition too invites the aspirant to transcend limited definitions. Buddha compares the body to a thing ablaze.[33] Before becoming the Buddha (Awakened One) young Siddhartha Gautama is said to have had a transformative encounter with a sick man, an old man and a corpse, along with a renunciate.[34] He didn't simply pity the broken men: he knew this would happen to him and was the universal fate of human beings. This triggered his quest for the wisdom of mind and compassion of heart needed to end suffering. Some form of this archetypal journey is launched for many when our bodies break down: we too realize the truth of sickness, old age, mortality—it is happening to us. We too may inaugurate a quest for what abides or transcends in the face of this suffering.

For Buddhists this could be said to be our connectedness *to the entirety of the universe*—something perhaps similar, though different language is used—to that transcendent identity evoked by the *jnana yogi*.

Again, I will refer to this as the "'*I am*' possibility," and in two ways. First, illness opens up to us the possibility of an "I am" that outruns the broken body. This may, but need not, take religious or spiritual form, according to the beliefs of the individual patient. Second, any experience of this "*I am*" itself re-possibilizes the world. Again, Wendell writes of how one "expands the possibilities of experience beyond the miseries and limitations of the body." Carel values the "freedom of thought, imagination, emotion and intellect" that one finds even in the face of extreme physical incapacities. Ram Dass speaks of the peace and security that comes from identifying with Awareness.

In using the musical conceit of first, second, and third movements, I am not implying that these are stages that necessarily unfold in sequence, nor that one is "higher," or "further along" than the others. Keeping the verbal continuity—(i) *impossibility*, (ii) *"I'm-possibility"* and (iii) *"the 'I am' possibility"*—is actually meant to suggest how these moments are born from one another, and reciprocally interpenetrate. Someone dealing with a serious impediment—for example, an amputated limb—may experience in the same day, perhaps even at the same time, (i) frustration with his/her broken body, felt as something limited and freakish (*impossibility*); (ii) a compensatory use of his/her other limbs to complete a task (*I'm-possibility*) and (iii) a sense that he/she is so much more than just this body anyway (the *'I am' possibility*). *Re-possibilizing* our life in such ways doesn't just negate the body-impossibles that challenge us, but includes and integrates them.

Coda

A musical coda may be both summative, and yet improvisatory and playful. It is in this spirit that I turn my focus, finally, not to Perlman's violin, but Perlman himself, his body, biography, even his name.

The story of Perlman's busted violin clearly takes on added resonance in the light of Perlman's own story of a childhood polio that left significant and irremediable marks. He has long lived with a "three-stringed violin" of a body, motorically impeded. Fortunately, he writes,

> My parents were very supportive in a normal way. It was never: "Oh, we have a child who has polio!" . . . With my childhood friends, the abnormal thing about me was not that I walked with crutches—they got used to that. The abnormal thing about me was that I had to practice three hours a day. The other kids thought that was crazy.[35]

For a time, Perlman did not wish others to focus on the disability, or how it might have shaped him as a musician. He was angered with headlines like "Crippled Violinist Plays Concerto." Yet, in time, he came to own the disability as part of his identity and speak on disability-rights issues such as removing architectural barriers to access. He even refused to play in some venues because of access problems, whether for himself or the audience.[36] As mentioned earlier, re-possibilizing the world involves social cooperation. Credit is due to Perlman's parents, childhood friends, teachers, impresarios, concert hall architects and all who have helped make possible his genius. But credit is also due to Perlman himself. Refusing to be limited by his motor impairments, he accepted "the artist's task to find out how much music you can still make with what you have left."

J. S. Bach chose to use his own name as musical note sequence, B A C H (H is the German notation for what we would call B natural), in the last Contrapunctus theme of the *Art of the Fugue*, the piece he was working on in his final days. I beg the reader's indulgence while I end this piece with a brief theme based on Itzhak Perlman's name.

Itzhak, the Hebrew version of the English "Isaac," comes from the Biblical story of Sarah and Abraham. When they were both quite old, such that Sarah had "stopped having the periods of women," a divine emissary tells her that she is to have a child . . . and she laughs! " 'Now that I am withered, am I to have enjoyment—with my husband so old?' . . . The Lord asked, 'Why did Sarah laugh . . . Is anything too wondrous for the Lord?' " (Gen 18:10–14).[37]

When Sarah does in fact bear a child, she names him Itzhak, which in Hebrew means "to laugh." Sarah says, "God has brought me laughter; everyone who hears will laugh with me" (Gen 21:6). In fact, Itzhak (Isaac) becomes the progenitor of the Jewish people, a people based, so speak, on a divine joke.

We see here, in playful register, the themes of this chapter woven together. There is a seeming physical *impossibility*—new birth from withered bodies? Yet Sarah and Abraham claim the stance that "*I'm-possibility*." In later life they take on new names, move to a new land, and even have a miraculous child. Great music thus comes from three-stringed violins. Finally, and obviously, the story speaks to a transcendent dimension, what I have termed "*the 'I am' possibility*." Sarah and Abraham hear a call of the Divine, whose name is "I am that I am" (Ex 3:14), allowing those who hear and follow to transcend limitations.

So much for the name "Itzhak," but what of "Perlman?" Well, our artist is something of a Pe(a)rl-man. A pearl is formed when a parasite or other irritant lodges within an oyster, mussel or clam, leading it to deposit concentric layers of mineral. This embodied adaptation creates a thing of jewel-like beauty. So, too, may something comparable happen when we face serious illness, injury or impairment. For Itzhak Perlman, in the story wherein his violin string broke—perhaps even earlier as a child when some of his body broke—a precious pearl was the product of adaptive struggle.

This can serve as an inspiration, but hopefully not a reprimand, for those struggling with illness and impairment. Few of us retune the violin as quickly and expertly as Perlman is portrayed as doing that night. In fact, the demand for high-performance recovery—to always be seeing the good, remaining cheerful and energetic despite pain and disability, to make instant lemonade out of all of life's lemons—can create unneeded pressures and suffering for the ill. For many, when a string breaks, the best we can do is take time to grieve, and then pick up the fiddle and somehow keep sawing away. Over the course of time we may hit on some beautiful notes. Or at least, like Sarah, we may have a laugh as we struggle along with disruptions and new births.

Notes

1. Jack Riemer, "Perlman Makes His Music the Hard Way," *Houston Chronicle* (February 10, 2001), www.chron.com/life/houston-belief/article/Perlman-makes-his-music-the-hard-way-2009719.php
2. For a summary of possible problems with this story, see www.snopes.com/fact-check/three-strings-and-youre-outre/ I am also indebted to Joseph Straus, an exceptional musician, music theorist and author in disability studies, for pointing out musical questions about this account, and for his many other helpful suggestions with this piece.
3. See, for example, Arthur Kleinman, *The Illness Narratives: Suffering, Healing and the Human Condition* (New York: Basic Books, 1988); S. Kay Toombs, *The Meaning of Illness: A Phenomenological Account of the Different Perspectives of Physician and Patient* (Dordrecht: Kluwer Academic Publishers, 1992); Thomas Couser, *Recovering Bodies: Illness, Disability, and Life Writing* (Madison: University of Wisconsin Press, 1997).
4. Ram Dass, *Still Here: Embracing Aging, Changing, and Dying* (New York: Riverhead Books, 2001), 71.
5. Susan Wendell, *The Rejected Body: Feminist Reflections on Disability* (New York: Routledge, 1996).
6. Drew Leder, *Spiritual Passages: Embracing Life's Sacred Journey* (New York: Penguin Books and Tarcher, 1997).
7. Wendell, *Rejected Body*, 20.
8. Paul Kalanithi, *When Breath Becomes Air* (New York: Random House, 2016), 140–1.
9. Kevin Aho, "Notes from a Heart Attack," in *Phenomenology of the Broken Body*, ed. Dahl, Falke, and Eriksen (London: Routledge).
10. Drew Leder, *The Distressed Body: Rethinking Illness, Imprisonment and Healing* (Chicago; University of Chicago Press, 2016), 13–23.
11. Aho, "Notes from a Heart Attack," p. 193.
12. Wendell, *Rejected Body*, 27.
13. Nancy Sherman, *The Untold War: Inside the Hearts, Minds, and Souls of Our Soldiers* (New York: W. W. Norton and Company, 2010), 197.
14. Ibid., *Untold War*, 203.
15. Wendell, *Rejected Body*, 86.
16. Havi Carel, *Illness: The Cry of the Flesh* (New York and London: Rougledge, 2014), 65.
17. Aho, "Notes from a Heart Attack," p. 196.
18. Arthur Frank, *At the Will of the Body* (Boston: Houghton Mifflin, 1991), 58.
19. Jean-Paul Sartre, *Being and Nothingness*, trans. Hazel E. Barnes (New York: Washington Square Press, 1993).
20. Maurice Merleau-Ponty, *Phenomenology of Perception*, trans. Donald A. Landes (London: Routledge, 2012), 139.
21. Carel, *Illness*, 97.
22. Merleau-Ponty, *Phenomenology of Perception*, 144.
23. Ram Dass, *Still Here*, 71.
24. Carel, *Illness*, 146.
25. Ram Dass, *Still Here*, 194.

26. Ibid., 184–204.
27. Aho, "Notes from a Heart Attack," p. 200.
28. Leder, *Distressed Body*, 24–41.
29. Carel, *Illness*, 83.
30. Wendell, *Rejected Body*, 177–8.
31. Ram Dass, *Still Here*, 201.
32. Ibid., 194.
33. Lucien Stryk, ed. *World of the Buddha* (New York: Grove Press, 1968), 53–5.
34. Ibid., 25–6.
35. Cited in Joseph N. Straus, *Extraordinary Measures: Disability in Music* (Oxford and New York: Oxford University Press, 2011), 143.
36. Ibid., 143–5.
37. Translations from W. Gunther Plaut, ed. *The Torah: A Modern Commentary* (New York: Union of American Hebrew Congregations, 1981).

11 Notes from a Heart Attack

A Phenomenology of an Altered Body

Kevin Aho

Introduction

It was a beautiful December day in Southwest Florida—a Saturday, bright sun, blue sky, low humidity. I planned a 60-mile solo bike ride around the town of Estero, the Fort Myers Airport and the neighboring communities of Gateway, Treeline and Colonial. Finishing up the ride in around 3.5 hours, I climbed over the Estero Bridge to my house when I was suddenly overcome with nausea and lightheadedness. I slammed on the brakes, threw my bike to the ground and vomited all over the street. Confused and thinking I had food poisoning or had overdone it on the ride, I slowly pedaled back home. Then the chest pain came as a dull, persistent ache. I called my girlfriend, telling her that I was having some trouble. She said it sounded like I was having a heart attack. I dismissed it. "No, I'm just hungry and dehydrated and need to take a shower." She raced to my house, convinced me to go to the Emergency Room as the dizziness deepened. A quick EKG ushered me into a suite of scurrying doctors and nurses who were already preparing for the surgery. All I heard from the din was "Massive Heart Attack . . . LAD blocked . . . Code Stent . . . Code Stent!" *This was the beginning.*

A week after the heart attack was Christmas Day, and I was feeling much better. Always physically active, my girlfriend and I now committed to a heart healthy diet and were preparing meals at home for the first time in our relationship. I had walked five miles during the day and took a hot bath in the evening. I was returning to my old self again, more quickly than I could have imagined. I would be healthy and whole before the start of the spring semester. The next day, I started out on my morning walk but only got to the end of the driveway. My right calf felt tight and achy and my toes were numb. I came back to the house with a grim "Something's wrong." We rushed back to the hospital where I received an ultrasound on my leg and sure enough, a blood clot in a major artery. Three days of treatment with a vascular surgeon, angiograms to examine the clot and various tubes inserted through my left groin down to my right calf (the right groin couldn't be used as that was the side they had gone up to place the stent). The surgeon was unable to suck the clot out with a tiny vacuum, so he opted for an aggressive intravenous clot buster treatment combined with high doses of a blood thinner. I was unable to eat or stand for the duration. Every hour, the nurse would measure the size of my calf to see if blood was flowing, and each hour I was gripped with terror that it was getting larger or that the pulse in my right foot was getting weaker. Each night was a din of buzzers, beeps, blood tests and vital sign checks. I slept in fits and starts. *This was the beginning.*

I was finally released from intensive care after the clot buster medication had done its work, and I was able to move to my own hospital room for observation. Having my own bathroom, a window and the ability to eat solid food was succor. The diagnosis was that a clot in my heart had been discharged during the heart attack, and that I would need to be on a battery of blood thinners to prevent future clots from forming. This was bad but not terrible. I could take the medications, go to cardiac rehabilitation and get back to my normal routine in no time. On the second night of observation, an alarm and flashing red light erupted from the heart monitor hanging on the wall. A 30 second burst of ventricular tachycardia. Now things got complicated. The next morning the cardiologist warned that I might be susceptible to sudden cardiac death and needed to see an electro-cardiologist to determine the extent of the problem. But the test, involving electrodes placed on the heart to induce tachycardia, could not be done because of concern that it would dislodge the clot in my heart and lead to a possible stroke. The solution was to wear a portable defibrillator (or life-vest) for two months as the clot dissolved and then decide whether or not to implant a defibrillator in my chest. I was stricken with the reality that the heart attack and the various complications had left me with a profoundly altered body. I am now forced to confront the possibility that reclaiming my former self is not an option. *This is the beginning.*

What Is an Altered Body?

To have a massive heart attack at the age of 48 at, what I thought was, the prime of my life has been nothing less than world shattering. The visceral awareness of the frailty and vulnerability of my body and the recognition of my fundamental dependency on others have trashed the illusions of my own strength and autonomy. But the experience has also concretized insights in the phenomenology of the body that I have been reading and writing about for years, revealing a now intimate understanding of critical illness and the limitations of mainstream biomedicine. As is well known, the signature contribution of phenomenology when it comes to an understanding of body ailments is the way in which it brackets out or suspends the detached third-person perspective of scientific medicine, focusing instead on the first-person perspective, that is, the lived experience of the suffering individual. To this end, phenomenology is not concerned with identifying the causal explanation (*Erklärung*) of my heart attack through the technical use of echocardiograms, X-rays or blood tests. Instead, the aim is to arrive at a descriptive understanding (*Verstehen*) of *what it means* and *what it feels like* to experience a heart attack. Thus, the phenomenologist is concerned with the experience of "illness" *as it is lived*, to understand it, make sense of it and give it meaning.[1] By closely attending to lived experience in this way, the aim is to return to the things themselves, to the structures of meaning that constitute our subjectivity and the way these structures are altered and break down in the course of critical illness. This requires an alternative to the way in which mainstream biomedicine interprets the body.

What became clear in the aftermath of the heart attack is that the physicians and surgeons rarely encountered me as a person. I was reduced to numerical data, to numbers and sets of numbers revealed through various diagnostic instruments. I was a 45–50% ejection fraction, an INR of 5.7, an S/P sustained 83 VTACH or an ST > 0.15V. For every procedure or blood test, I was not Kevin Aho, a middle-aged philosophy professor, with parents who were worried about him, with two loving brothers, a girlfriend and a career that he cared deeply about, who was now struggling and overwhelmed in the face of a collapsing world. There was little attempt to listen to the person behind the symptoms. I was simply a corporeal thing, an object of measurement. In the German phenomenological tradition, there is a sharp distinction between the corporeal body (*Körper*) and the lived body (*Leib*).[2] The corporeal body is encountered from the perspective of theoretical detachment. This is the account of the body we inherit from Cartesian and Newtonian science where bodies are defined in terms of having a material composition that is measurable and subject to the laws of nature; occupying a spatial location and having determined boundaries; and being viewed dispassionately as an "object" under the gaze of the cognizing "subject."[3] Regarded from the perspective of *Körper* in the hospital, I felt like valueless matter whose functions were being ceaselessly measured and re-measured. This pattern of measuring blood pressure, temperature, pulse and respiration rates took place every four hours for 12 straight days, pulling me out of deep sleep every night at 2 am. Each time I was poked and prodded, my rest and recovery were interrupted, leaving me frustrated and exhausted. It seemed as if the health care professionals saw my body as something *I have* rather than who *I am*. They misunderstood, in Martin Heidegger's words, that "we do not 'have' a body; rather, *we 'are' bodily . . . we are somebody who is alive*."[4] Moreover, the numbers gathered from the measurements every four hours were not impersonal facts. They meant something to me, pointing toward what kind of life I can live, whether or not I can ride my bike again or travel overseas or flourish in my profession.

From the standpoint of my own experience, then, the body is not a corporeal thing that the physician examines from a standpoint of cool detachment. It is, rather, a lived body (*Leib*). Related to the German words for "life" (*Leben*) and "experience" (*Erlebnis*), the lived body refers to *my body* and the panoply of experiences, feeling, meanings and interpretations that belong to me. As Gabriel Marcel writes, "what I feel is indissolubly linked to the fact that my body is *my body*, not just one body among others. . . . Nobody who is not inside my skin can know what I feel."[5] But understanding the lived body requires more than being attentive to the first-person experiences of the patient as they are lived. Phenomenologists also acknowledge the importance of what Maurice Merleau-Ponty refers to as the "body-schema" (*schéma corporel*), that is, the mediating activity of the lived body, of my physiology and the perceptual and postural systems that pre-reflectively orient and situate me in the world. When I am healthy, these sensory-motor systems remain hidden in the background, opening up the practical horizon or space of my life and seamlessly coordinating my movements and position in relation to the vertical (up and down), the horizontal (front and back) and other axes.[6] The mediating activity of the lived body constitutes a tacit sense of, what Merleau-Ponty calls, the "I can" as I engage in the familiar habits and patterns of everyday life.[7] With my heart attack, the transparent functions of the body schema collapsed, and the corporeal body emerged out of its hiddenness. Hans-Georg Gadamer captures this experience when he writes, "[We] know only too well how illness can make us insistently aware of our bodily nature by creating a disturbance in something which normally, in its very freedom, almost completely escapes our attention."[8] With the disturbance of the body schema, I became aware of my body as an object, as something foreign and strange. Every pinch in my chest, every constricted breath and skipped heartbeat pulled me away from the "I can" and injected doubt and worry into everything I did, resulting in a profound alteration to the structures of meaning that constitute who I am.

Spatial Wounds

From the perspective of lived experience, space is rarely encountered as a three-dimensional coordinate system that contains physical objects. It is, rather, the concrete setting of my life, the "*wherein* [that I] live."[9] When the phenomenologist refers to human existence as "being-in-the-world," then, he or she is not referring to the idea of spatial inclusion as if I were residing inside a worldly container. He or she is referring to the way I am always already involved in the world, and this involvement opens up and constitutes space as the experiential horizon or field of my life. As Heidegger writes: "The human being makes space for himself. He allows space to be. An example: When I move, the horizon recedes. The human being moves with a horizon."[10] Before my heart attack, this horizon was expansive and was constituted effortlessly by simply moving or "bodying forth" (*Leiben*) through the world.[11] Taken for granted activities like driving, walking, negotiating stairs, giving a lecture or picking up and handling heavy objects were performed in a smooth and transparent way, and this held open and broadened the space of my everyday concerns. In the aftermath of the heart attack, this space quickly closed in on me. I vividly recall the yellow footprints painted on the floor of my hospital room that ran from the bed to the bathroom. There were four footprints with a sign on the door that read, "Walking helps your recovery." When I asked the nurse what this meant, she said it refers to the importance of walking to the bathroom. In the span of a few days, the horizon of my life had shrunk from 60-mile bike rides to four steps. In many ways, I felt as if I was moving from *Leib* to *Körper*, from the effortless activity of "making space" to merely "occupying space" as a corporeal thing.[12]

This experience of spatial collapse didn't end when I was discharged from the hospital. There were constant reminders of my new limits. A kind of agoraphobia set in. The neighborhood streets in the evening seemed threatening, especially when my house was no longer in view. A drive to the grocery store or to work took on frightening implications. And flying to a new city for a conference or to visit friends and family seemed impossible. The collapse was so acute that I asked my younger brother to come to stay with me for a few days when my girlfriend was out of town for work, fearing that I'd be unable to leave the house. There were also more subtle reminders. The bike rack on the roof of my car continued to haunt me, indicating the loss of strength and independence. Social media feeds on Facebook and Instagram were filled with photos of smiling faces, dinners with friends, exotic locations and ski outings, all suggesting how narrow and constricted the space of my life had become. When looking out my office windows, I watch young students playing Frisbee on the lawn, riding their skateboards, and laughing as if to remind me of my new frailty. Even the architectural design of the built environment seemed like a spatial betrayal. The stairs to my office or to my car in the parking garage left me short of breath, the long boardwalk, the distance to the library were all signs, were all whispering of my weakness. The "I can" that opened up my world, that I had effortlessly embodied just a few weeks earlier, had become a debilitating "I can't."[13]

Temporal Wounds

As a method, phenomenology famously rejects the traditional view of temporality, of clock-time as a linear sequence of measurable now points. Time, from the perspective of my own existence, is the finite span of my life, one that stretches forward and backward, simultaneously opening up possibilities in the future and constrained by the limitations of the past. Heidegger refers to this temporal movement in terms of "thrown projection" (*geworfen Entwurf*), conveying the idea that existence is inescapably thrown into a past, into a socio-historical situation that we did not choose, and it is this situation that opens up possibilities that we can project for ourselves in the future.[14] For Heidegger, then, human existence is both "ahead of itself" (future) and "what it was" (past), and it is the unity and coherence of this temporal structure that constitutes the site of meaning—the *Da* of Dasein—that allows us to understand the world and make sense of who we are. Before my heart attack, the future was expansive and filled with worthwhile projects shaped by a past that was grounded in physical strength, self-confidence and a generally upbeat temperament. I was running toward a future of continued health and professional success, of travel and writing, of cultivating new friendships and sustaining old ones. There was a sense of integrity and solidity to who I was because the meanings and interpretations I projected for myself were coherent and bound together with my past. My illness shattered the illusion of this temporal unity, exposing an arrested future of medical tests, of hospital visits, of medications and monitoring.

When the future collapsed in this way, the resources of the past that I had relied on to create my identity no longer held, and I was left to confront the ultimate question, "Who am I?" I was no longer the productive professor, the cyclist, or the skier. These former identities no longer made sense to me in the wake of the heart attack. My future projects shifted from writing a new book or traveling to France with my girlfriend to walking to the bathroom without losing my breath. I felt as if the future and the past were closing in me, as if I were trapped in a meaningless present, left to the moment-to-moment rituals of taking medication, checking my blood pressure and pulse and arranging the next doctor visit. In *Being and Time*, Heidegger refers to this kind of experience in terms of "anxiety" (*Angst*). Anxiety discloses a future that no longer holds open a range of meaningful possibilities. It reveals a future that is uncanny, stripped of value and meaning. In this state, writes Heidegger, "the world collapses into itself; [it] has the character of completely lacking significance."[15]

As an experience of world collapse, anxiety is not to be viewed as an emotion. Emotions are short-lived, episodic experiences that have determinate causes and are generally directed at specific objects in the world. For Heidegger, anxiety is a "mood" (*Stimmung*), an atmospheric or global affect that is not directed at worldly objects or events but to the world *as a whole*. Originally referring to the tuning of a musical instrument, the German word *Stimmung* captures the sense of an affect that "tunes" us to the world in a particular way and envelops everything. It is, as Heidegger writes, "already there . . . like an *atmosphere* in which we first immerse ourselves in each case and which then attunes us through and through."[16] As a mood, a diffuse anxiety colored the atmosphere of my life. Things that I used to take pleasure in—watching a good movie, reading *The New York Times* on Sunday or enjoying a nice meal—now revealed themselves as affectively empty and flat. The plot line and dialogue of the movie was unintelligible, the words in the newspaper article were confusing and disjointed, and food was bland and tasteless. Heidegger claims that moods are disclosive; they allow things to matter to us in the ways that they do and make manifest "how one is, and how one is faring."[17] If human existence is the "mooded" (*stimmungsmässigen*) site of meaning through which things reveal themselves, then my post-heart attack anxiety revealed a bleached out world, where the things that I encountered and handled everyday were, with few exceptions, affectively meaningless. As a result, the activities that used to stand out as significant for me, the very things I drew on to construct a unified and coherent self-interpretation, imploded, and I was confronted with a kind of "dying" (*Sterben*), what Heidegger calls "the possibility of the *im-possibility* of existence."[18] To be human requires the capacity to understand and make sense of the world. As Heidegger says, "to *exist* is essentially . . . to *understand*."[19] The heart attack disrupted the future-directed understanding that I had of my life, undermined the constitutive meanings of my past, and left me in a state of limbo, unable *to be*.

Hermeneutic Wounds

In the tradition of existential phenomenology, the idea that we create or fashion our identities against the past that we've been thrown into is famously captured in Jean-Paul Sartre's claim that "existence precedes essence." This means that, unlike other animals, humans have no pre-given nature or essence that determines who we are. I cannot be viewed as a mere corporeal object, because I *exist*. That is, I have the capacity to make choices, to interpret, understand and give meaning to my corporeal givenness. This suggests that human existence is fundamentally hermeneutic; that I am a *self-making* or *self-fashioning* being that exists only in the meanings and interpretations I create for myself. This is why the words of the physician are so powerful. The diagnosis of "ischemic cardiomyopathy," "myocardial infarction" or "coronary artery disease" are not merely diagnostic labels; they are symbols that reflect my situation and that I now must draw on to create a new identity. Viewing me from the lens of *Körper*, the physicians and surgeons seemed largely oblivious to the power of their own words. After I was admitted back into intensive care with a blood clot, I remember an especially callous doctor telling me "It may require surgery, but who knows?" I remember hanging on these words and replying angrily, "You should know! You're the damn doctor." After my run of ventricular tachycardia, when a cardiologist entered my hospital room to say that I am now at risk of sudden cardiac death and would probably need to have a defibrillator implanted in my chest, the words "sudden death" and "defibrillator" left me reeling. Before I had a chance to comprehend this news, to ask, "What does this mean for me?" and "How am I going to live?" he was already gone, walking briskly down the hall. A few days later, another cardiologist came to my room and said I would need to wear a life-vest for the next two months. Again, the very word "life-vest" darkened my future, dimming my capacity for self-creation.

Creating a unified and coherent sense of self requires, what Heidegger calls, a "for-the-sake-of-which" (*das Worumwillen*), referring to a tacit recognition of our own "futurity" (*zukünftig*), of where our life is heading, rooted as it is in the discursive resources of the past. The medical terms that captured my diagnosis closed down this sense of futurity, but they also corrupted my past. Experiences of strength, confidence and vitality that were so much a part of my self-interpretation before the heart attack suddenly seemed foreign, as if they belonged to someone else, and the future revealed itself as a horizon of weakness and futility, one that was confined to the medical industrial complex. And the detached discourse of the health care professionals only exacerbated the experience, turning me into a passive object and stripping away any sense of agency. Physicians would often discuss my condition from the third-person perspective in front of me, as if I weren't even in the room. "Patient was treated with beta blockers and an ace-inhibitor." "Patient's pulse is brady." "Patient was transferred to the cath lab." The focus was on treatment rather than healing, on *what* was being done to my corporeal body rather than *who* was living or undergoing the treatments. When I would ask for clarification regarding the significance of these various treatments in an attempt to understand and make sense of them, I was dismissed. It became clear that it was not even the physicians but the technology itself that was the authority. The physicians deferred not to me, not to how I felt, but to the data transmitted on the heart monitor, the X-ray, or the ultrasound. Any attempt on my part to intervene in the sheer instrumentality of the proceedings appeared to be viewed as a disruption in the delivery of care. The overall impact of this objectification was a feeling of profound helplessness, that I was no longer a participant in my own existence.

Inter-corporeal Wounds

Phenomenology is well known for its rejection of the modern idea of the self as an encapsulated, atomistic individual, arguing that human existence is structurally intersubjective and relational; it is always embedded in a common world and open and receptive to the lives, bodies and expressions of others. This means, in Heidegger's words, that the human being "is never just an object which is present-at-hand, [and] it is certainly not a self-contained object."[20] As a way of being-in-the-world, human existence is *ec-static* in the literal sense of standing (*stasis*) outside (*ex*) of ourselves to the extent that we are already bound up and involved in a wider context of shared meanings and values. This is why Heidegger claims that, " 'being-in-the-world' is always 'being-there-with-others.' "[21] When healthy, I was absorbed in the relational flow of the world, seamlessly engaging with public life and the various social roles and practices that fortified my sense of self. Involved in this way, I was invisible, a "they-self" (*Man-selbst*) who could easily vanish into the web of social relations.[22] But illness disturbed this synchronous flow and made me visible. I was now too slow and clumsy on the staircase, unable to keep up on brisk walks to the library or finish a three-hour lecture in the afternoon. Without the ability to disappear into ordinary social situations, I felt as if I was reduced, in Sartre's words, to a "body-for-others" (*corps pour autrui*), an object under the gaze of the healthy and normal.

The experience of myself as a body-for-others was magnified when I had to start wearing the life-vest. Although the external defibrillator was largely concealed under my shirt, it was attached to a camera-sized box at my hip, with a black cord running up my side. It was unmistakably a medical device, and I felt the stigma. When colleagues would approach me, they would glance at the device and ask with concern, "How are you doing?" and "Is everything alright?" questions that enflamed my sense of brokenness. What was especially disturbing is that some colleagues, whom I considered friends before the illness, appeared to avoid me altogether or would simply smile and scurry away, perhaps uncomfortable with what I represented, shattered health, vulnerability and a reminder of death. Others, after asking about my condition, would redirect the conversations back to themselves, describing their excellent cholesterol levels, lack of heart disease in the family, fitness regimen or the value of their plant-based or vegan diet. Through all of this, the gaze of the other branded me as ill, as an outsider, as someone who disrupted their tacit harmony. And all of this made me acutely aware of my altered body. I began to internalize the judgment of others, to see myself as others saw me. This gave me a newfound appreciation for Sartre's account of "the look" (*le regard*) in *Being and Nothingness*, when he writes, "The other's look fashions my body in its nakedness, causes it to be born, sculpts it, produces it as it is. . . . The other holds a secret—the secret of what I am."[23] In attempts to avoid "the look," I would try to disguise my condition. I would present myself as healthy, upbeat and strong when colleagues or students came to the office; I would hide my medical device on my lap or carry it over my shoulder like a fashion accessory; I would make up excuses about why I couldn't attend a meeting or event; I would pretend to take a phone call if I felt tired or overwhelmed. All of this was done in an effort to disappear, to vanish back into the flow of everydayness.

Healing the Altered Body

Heidegger refers to illness in terms of "a loss of freedom, [and] a constriction of the possibility of living."[24] As we have seen, this constriction can take a number of forms, from spatial and temporal collapse, to constrictions in our narrative possibilities and relational ways of being. Each of these represents a disruption or breakdown in the structures of meaning that constitute our subjectivity. Healing, on this view, involves becoming receptive and open to the world once again, to projecting new possibilities and meanings against the background of the illness. For me, this process is taking place slowly, beginning with simple movements. From being bedridden, to being able to stand, to walking down the hospital corridor a few days later, to walking around the block, to walking several miles, these incremental expressions of motility began to create a new foundation for my experience of lived space. A horizon that consisted of a hospital bed for weeks began to gradually widen and expand. Instead of just *taking up* space as a corporeal object, my capacity for movement allowed me to *make space* once again. When I was cleared to drive, when I walked across campus to my office for the first time, when I taught my afternoon seminar, when I made love again, the space of concerns continued to broaden. But this spatial broadening is still tentative. There is a persistent fear of another incident, a shock from the defibrillator, a pain in the chest or a blood clot that will send me back to the constricted confines of the illness.

Disruption of the temporal structure of my existence has made it difficult to project myself toward a meaningful future. In the immediate aftermath of the heart attack, I was not only absorbed in the moment-to-moment management of my own limitations and despair, but it was also unclear to me how I could draw on the discursive resources of my past. It was as if the heart attack had established a new starting point, a new foundation for my life going forward. Any meaningful identity that I could envision was now constrained by the illness. And I would have to work against the background of these constraints to create myself again. Again, this process of reopening the future has been gradual. I've had to slough off identities that no longer hold for me. The most obvious is letting go of the sense of myself as someone who is fit, athletic and strong. Physical fitness had been integral to who I am since I was a teenager, and provided a decades-long release from symptoms of chronic stress and anxiety. Running, cycling, lifting weights at the gym were all like therapy. They calmed me down, released endorphins and fortified my confidence and feeling of well-being. With the heart attack as my new birthplace, the meanings of fitness have shifted dramatically. At least for the time being, I have to make-do with slowness and being patient, with walking, stretching and meditation, letting go of the past and acknowledging that there are many different ways to interpret the idea of strength.

My identity as a professor has also changed. I had long fashioned myself a productive scholar and an energetic teacher and colleague, but that energy and enthusiasm is diminished both by the illness as well as by the various heart and arrhythmia medications that make me light-headed and easily fatigued. Long days at the office or marathon writing sessions are difficult, and I have been forced to say "no" to projects and trips that I would normally jump on. This has all created tentativeness about my future as a professor. It feels uncertain, filled with what-ifs. Will I be able to travel with an implanted defibrillator? How will I cope if or when the device shocks me? What will my next echocardiogram reveal about the health of my heart? How will this affect my ability to do my job? Much of my energy is devoted to managing this kind apprehension, a fear that narrows and constricts the possibilities available to me. But with each day at the office, with every hour I spend in front of my students, with each word I type for a new manuscript, the future opens up a little more. In the process, I am recognizing that I have to be flexible, that I cannot cling to my former self-interpretation. Heidegger explores this notion of ontological flexibility in his discussion of "authenticity" (*Eigentlichkeit*) in *Being and Time*. To be authentic, for Heidegger, is to be steadfastly open and receptive to the possibility of death, that is, to the contingency and impermanence of one's own being (or identity). And this openness demands a willingness to be flexible, to let go of identities that are no longer livable. For this reason, writes Heidegger, one "cannot become rigid as regards the situation, but must understand that [authenticity] . . . must be *held open* and free for the current factical possibility."[25] The illness opened me up to the structural vulnerability of my being, and the path forward requires the capacity to give up on a former self that I am still tenaciously clinging to. Thus, the possibility for authenticity "lies in *giving itself up*, and [shattering] all one's tenaciousness to whatever existence one has reached."[26]

The pain of letting go or giving up on my former self was not made any easier by the physicians who largely seemed oblivious to my experience, confined as they were in the objectifying discourse of biomedical science. But there were some in the hospital, fellow sufferers, who listened to me, affirming and accepting my experience. Their attentiveness to me as a lived body rather than a corporeal thing created a new skein of meanings that have helped me refashion my life-story, with my illness now serving as a foundational chapter. There was a particularly dark moment in the hospital when my cardiologist had diagnosed ventricular tachycardia, the possibility of sudden cardiac death and the need for a defibrillator. A nursing assistant came in a few minutes later and sang a negro spiritual to lift my mood and let me touch her scar from her own pacemaker. She talked openly about wanting to commit suicide earlier in her life, of her congenital heart trouble and getting her pacemaker at 40. I was overcome by her compassion, by her capacity to see me as someone who was in pain. Then there was the nurse from the Intensive Care Unit, who confessed to his own cardiac issues from atrial fibrillation and spoke honestly about his anxiety and how he would have to give himself positive affirmations in front of the mirror in the morning in order to leave the house and go to work. Another nurse stopped by my room after a procedure, saw that I was suffering, and disclosed that he had cystic fibrosis and probably not long to live, but that his life was still worth living. And there was a young nurse, a former student of mine, who quietly told me one evening that she entered the healing professions to confront the pain of losing her father to colon cancer.

Each of these disclosures provided the recognition I desperately needed and offered a discursive context that I could draw on to express and make sense of my shattered identity. From these encounters, the interpretive structure of my existence remained intact, and I was, against the backdrop of my illness, able to envision a future that could still be meaningful and fulfilling. This highlights the importance of situating and acknowledging the existential suffering of the patient as fundamental to healing. The fact that those who listened to and affirmed my experience were mostly nurses (or nursing assistants) reveals something about the instrumental and transactional nature of modern doctoring, where the aim is not to listen and tend to a particular person but to manage and treat malfunctioning body parts. As Gadamer writes, "instead of learning to look for illness in the eyes of the patient or to listen for it in the patient's voice [the doctor] tries to read off the data provided by technologically sophisticated instruments."[27] Many physicians, including Dr. Bernard Lown, the cardiologist and inventor of the very defibrillator that I may soon be implanted with, have critiqued the industrialization of their profession for this reason. Today, writes Lown, "healing is replaced by treating, caring is supplanted by managing, and the art of listening is taken over by technological procedures."[28] For me, without others listening to and caring about my situation there would be no healing and, consequently, no way to project a meaningful path forward.

This aspect of healing, of being recognized and accepted as a person who was suffering, exposed deeper layers of the structure of intercorporeality or being-with-others, revealing the fundamental vulnerability at the core of the human condition. When I was healthy and caught up in the flow of everyday life this vulnerability remained largely closed off from me. But the heart attack cracked me open. I was suddenly overflowing, not just with anxiety but with love and compassion. I called my brothers weeping, shortly after the angioplasty, telling them how much I cared about them and how thankful I was to have them in my life. My girlfriend became my fiancée in the Intensive Care Unit after my blood clot. I felt as if I were seeing her for the first time, with fresh eyes, as the beautiful, courageous and tender being that she is. My parents, whom I spoke to everyday, revealed themselves in all of their generosity and devotion to their broken son. The people in my life that I ordinarily took for granted became luminous and fragile. And this extended beyond my family, partner and intimate friends, to colleagues, neighbors and even complete strangers at the supermarket or gas station. The masks came off, and I sensed how helpless and dependent we all are, and this made me love even more. The experience feels analogous to what Emmanuel Levinas, in *Totality and Infinity*, described as "the face" (*le visage*), when the other reveals him or herself not as a thing but as a pure expression of "nudity," "defenselessness" and "vulnerability."[29] This expression, for Levinas, not only illuminates *who we really are* beneath the stable crust of everyday social convention but issues a command or plea to take responsibility and care for each other.

This feeling of being cracked open has been the gift of the illness. As I struggle with the limitations my altered body, let go of a former self that is no longer livable, and work to refashion a new identity in the face of a precarious and restricted future, I am grateful for what the heart attack has taught me. I recognize now that I am not, and never have been, a masterful and autonomous subject, that I am fundamentally defenseless and dependent on others. And the recognition of our shared vulnerability is healing insofar as it binds us together in the wake of pain and loss, reminding us that we are not alone in our suffering. To this end, a phenomenology of illness not only allows us to see how the constitutive meaning-structures of our experience can break down but also how they can be rebuilt. The ways in which our existence is structured, by space and time, by our capacity to interpret and give meaning to the world, and by our intercorporeality, are always vulnerable to collapse. Illness reminds us of this. But, insofar as our suffering is not just treated with medical technology but *heard*, *acknowledged* and *affirmed* by others, there is still a way forward; there is still joy in being alive.

Notes

1. Martin Heidegger, *Being and Time*, trans. E. Robinson and J. Macquarrie (New York: Harper & Row, 1962), §§ 32–4.
2. Kevin Aho, "The Body," in *The Bloomsbury Companion to Heidegger*, eds. F. Raffoul and E.S. Nelson (New York: Bloomsbury Press, 2013), 269–74.
3. Edmund Husserl, *The Crisis of European Sciences and Transcendental Philosophy*, trans. D. Caar (Evanston, IL: Northwestern University Press, 1970), §§ 35–42. See also Edmund Husserl, *Ideas Pertaining to Pure Phenomenology and to Phenomenological Philosophy, Second Book*, trans. R. Rojcewicz and A. Schuwer (Dordrecht, The Netherlands: Kluwer Academic Publishers, 1989), § 28 and § 62.
4. Heidegger, *Nietzsche, Volume 1*, trans. D. Krell (New York: Harper & Row, 1979), 99. Emphasis added.
5. Gabriel Marcel, *The Mystery of Being: Reflection and Mystery, Volume 1.* (South Bend, IN: Gateway Editions), 104.
6. Maurice Merleau-Ponty, *Phenomenology of Perception*, trans. D. Landes (London: Routledge, 2012), 100–3.

7. Ibid., 103. See also Kevin Aho, "Temporal Experience in Anxiety: Embodiment, Selfhood, and the Collapse of Meaning," *Phenomenology and Cognitive Science* (Spring 2018), doi.org/10.1007/s11097-018-9559-x

8. Hans-Georg Gadamer, *The Enigma of Health*, trans. J. Gaiger and N. Walker (Stanford, CA: Stanford University Press, 1996), 73.

9. Heidegger, *Being and Time*, 83.

10. Heidegger, *Zollikon Seminars*, trans. F. Mayr and R. Askay (Evanston, IL: Northwestern University Press, 2001), 16.

11. Ibid., 86.

12. K. Aho, "Existential Medicine: Heidegger and the Lessons from Zollikon," in *Existential Medicine: Essays on Health and Illness*, ed. K. Aho (London: Rowman & Littlefield International, 2018), xviii.

13. Iris Marion Young, *Throwing Like a Girl and Other Essays in Feminist Philosophy and Social Theory* (Bloomington, IN: Indiana University Press, 1990), 36.

14. Heidegger, *Being and Time*, 185.

15. Ibid., 231.

16. Martin Heidegger, *Fundamental Concepts of Metaphysics*, trans. W. McNeill and N. Walker (Bloomington, IN: Indiana University Press), 67. Emphasis added.

17. Heidegger, *Being and Time*, 173.

18. Ibid., 307.

19. Martin Heidegger, *Basic Problems of Phenomenology*, trans. A. Hofstadter (Bloomington, IN: Indiana University Press, 1982), 276, my emphasis.

20. Heidegger, *Zollikon Seminars*, 3.

21. Heidegger, *Being and Time*, 152.

22. Ibid., 164.

23. Jean-Paul Sartre, *Being and Nothingness*, trans. H. Barnes (New York: Washington Square Press, 1956), 475.

24. Heidegger, *Zollikon Seminars*, 157–8.

25. Heidegger, *Being and Time*, 307.

26. Ibid., 308.

27. Gadamer, *The Enigma of Health*, 98.

28. Bernard Lown, *The Lost Art of Healing* (New York: Random House, 1999), xiv.

29. Emmanuel Levinas, *Totality and Infinity: An Essay on Exteriority*, trans. A. Lingis (Pittsburgh, PA: Duquesne University Press, 1969), 199.

12 Broken Pregnancies
Assisted Reproductive Technology and Temporality

Talia Welsh

Introduction

Discussions around pregnancy, childbirth and childrearing have been famously absent in phenomenological philosophical discourse. However, this tradition has changed sharply with many articles and books appearing in the last decades directed toward exploring the topic of pregnancy, not just as one area phenomenology has missed, but as perhaps an area that may have relevance for phenomenology in general. This chapter thinks about the ways in which pregnancy interacts with the medical sciences in contemporary developed worlds not as an experience, but as a *possible* experience for all women. In this way, assisted technologies of pregnancy provide a very different horizon regarding pregnancy and make infertile bodies not appear as permanently broken, but rather as temporary problems that can be solved with the right medical intervention. In such a way, assisted pregnancies are an example of our contemporary experience of the medical field as an infinite horizon upon which any ailment one day will be cured. In such a world, brokenness always is temporary and an error to be corrected rather than one common feature of what it is to be human. A phenomenology of brokenness within assisted pregnancy can aid us in thinking about our contemporary experience of temporality in the face of a progressive medical science. The value of a progressive medical science is evident; however, it brings with it little room to think of oneself outside of it. The cures for brokenness entrench brokenness within us making much of our embodied experience something that is in search of betterment through technological and scientific assistance.

This chapter first discusses how phenomenologies of pregnancy illustrate the porous boundaries between self and other. I discuss my own experience with Artificial Reproductive Technology (ART) as an experience of brokenness. I situate ART within our general experience of living within a medicalized world where every ailment and dysfunction has the promise of being solved one day. ART removes pregnancy as a mysterious, inexpressible experience since pregnancy becomes thoroughly constituted by medical intervention, testing and monitoring. For women,

ART extends their capacities for birth into later years, but in so doing highlights the difference between male and female experiences of fertility and refuses the traditional closure of childbearing years. ART exposes how our bodies become states of possible constant repair and modification. Mourning a bodily difficulty is always deferred to the next procedure, the next trial leaving one in a state of limbo. I conclude by asking if this makes brokenness paradoxically more endemic to the human condition since the future promises that all bodies can be treated, improved and repaired.

Medical Technology, Infertility and IVF

Years ago, I wrote a book chapter on phenomenologies of pregnancy. As a phenomenologist and someone who had never been pregnant and, at the time, had no interest in ever being pregnant, I was interested in the idea that perhaps the unique experience of intersubjectivity in pregnancy had a kind of philosophical meaning that could not be conveyed without actually experiencing the event—a kind of philosophically relevant limit experience or at least one that demanded a different way of expressing oneself—such as one finds in Luce Irigaray's provocative writing. I wrote:

> It is possible that the content of any experience is always-already lost in the abstract philosophizing experience. The idea that pregnancy is a limit-experience might help to highlight the potential relevance of an experience that cannot be circumscribed by a traditional phenomenological inquiry, but that appears in its heart deeply wedded to lived-experience. As such, we can see the idea of pregnancy as a limit-experience as continuous with the spirit, if not the execution, of a Merleau-Pontian inspired embodiment theory and harmonious with the challenging descriptions of split-subject, earth-world-home, maternal flesh, and an expanded self outlined in phenomenologies of pregnancy.[1]

In a similar vein, emphasizing the radical otherness of pregnancy, Nicholas Smith writes that "The temporality of pregnancy and the bodily transformations of the woman carrying the foetus, which go together with some of the most profound psychic alternations a human can go through, all are particular to the experience of pregnancy."[2] Pregnancy presents a challenge that is hard to incorporate neatly into pre-existing models of what it is to be a subject, which has obvious ramifications for phenomenology. Certainly pregnancy is a way to highlight the fact that our embodiment emerges from an other, making our intersubjectivity both primary and primal. Iris Marion Young writes how attending to the phenomenology of pregnancy expands upon the work of existential phenomenology, continuing "the radical undermining of Cartesianism that these thinkers inaugurated," but she goes on to point out that pregnancy "also challenges their implicit assumption of a unified subject and sharp distinction between transcendence and immanence."[3] For Young, pregnancy dissolves the transparent unity of the self. To try and grasp language to express the alterity of pregnancy, Sheila Lintott discusses the idea of calling for a feminist conception of the sublime modeled on gestation and giving birth. She writes, "The indeterminacy between the gestating woman and the being she gestates is spatial, temporal, existential, and at times internal to the woman and the being she gestates and births."[4]

The fetus and the mother are neither the same nor simply two different beings, and this discussion can be extended into thinking about ways in which our social nature is much deeper and incapable of being cut apart from individual experience. Thus, many phenomenologies of pregnancy work on these fundamental themes in phenomenology of self and other, self and social world, consciousness and unconsciousness, and the extension of the body into the world. While obviously impacted by changes in culture and history, these explorations can be seen as expressing something general about human life and should be able to inform a variety of discrete experiences of pregnancy.

In such a way, phenomenologies of pregnancy often highlight the irreducibility of pregnant experience to a neutral or genderless phenomenology. In addition, they draw attention to the universality of all persons arising not just from birth, but from the uterine experience, and the veil of early infancy and childhood. Unlike dying, to which I may come close to fully aware, my pre-birth experience is hidden from me. Moreover, my sense of self was a developmental acquisition over time emerging from the obscurity of early childhood. Smith writes that the fetus limits our ability to speak of self-consciousness, but the same time it extends toward such capacities. "Pregnancy thus seems to bring us to the very limits of rationality, language, and self-consciousness, while still being connected to these since the child will eventually come to acquire these capabilities."[5]

Years later, I did have a child. From the other side, I was struck by having no discernable experiences of the sublime or a radical undermining of my sense of self-constitution. Unlike my anticipation having read numerous phenomenologies of pregnancy, I did not discover pregnancy to be a particularly philosophically relevant limit experience. It certainly was like no other experience, and in this sense lived up to the idea of something inexpressible except by experience, but it did not press the limits of my subjectivity *experientially*. Nor did giving birth present me with a deeper sense of intersubjectivity beyond what anyone so inclined could ponder. This may have been because I was already philosophically invested in numerous trains of thought that work against the Cartesian model of self-sufficient rationality inhabiting a mind that constitutes subjectivity, so I had nothing to be "broken" by the experience.

To have a biological child with my husband we used in vitro fertilization (IVF) to get pregnant. At extraordinary expense (in the US IVF typically is entirely out of pocket to the parents), we went through two IVF trials, the first in which three embryos were implanted who all did not make it two weeks and the second in which three were implanted, two survived until eight weeks, one died and my son David was the one who made it full-term. I think my lack of finding myself stretched beyond ego-centered sense-constitution is due to the conditions of using assisted reproductive technology (ART), where no part of my pre-pregnancy and pregnancy itself was separate from a highly technological and mediated medical experience.

ART and Brokenness

To not be able to get pregnant without medical assistance is brokenness either in the person who wishes to carry the child or in the partner who wants to conceive with the mother. However, this brokenness both mirrors and radically differs from other commonly considered experiences of needing medical assistance. In the growing phenomenological examination of illness, disease and medical care, a distinction is drawn between *illness* as the experience and *disease* as what would be possible to examine through the medical sciences.[6] One might have a disease for which one has no experiential correlates, such as a slow-growing prostate cancer that is only known through testing. One might have an illness that presents no clear markers of a disease such fibromyalgia, known only through symptoms and the treatments looking to alleviate symptoms rather than being able to cure. Or one might have an illness for which no diagnosis is (yet) provided, struggling to express and have the medical community take one's experience seriously. While often disease and illness are conjoined, the value of having two categories to examine allows for the rich and varied ways in which individuals experience diseases and to help provide better and more caring treatments that take into account not just the body as a thing, but the body as a living person. It also helps better understand diseases that are greatly impacted or even caused by experiences such as stress and gets out of old-fashioned and unhelpful ideas of thinking there are "mental" states that are illusory in the individual and "physical" ones—i.e. those that can be "seen" in some manner that are real.

Brokenness spans the interconnections between illness and disease. For instance, it can be part of one's experience of illness or even something that is posited upon the individual as a disease but not experienced as such, such as in the case of disabilities that within a community of "disabled" persons are not considered limitations, such as in deaf communities that reject the idea that deafness is an inherent problem. Or one can experience brokenness because of the anticipation of the progress of a disease.

Bjørn Hofmann and Fredrik Svenaeus point out that the existence of ART can make an experience shift from one where one perceives oneself as unfortunate or unhappy to ill, "For example, assistive reproductive technologies have redefined the experience of childlessness from being faith or bad luck to be something to alter and treat by making it a disease (infertility)."[7] This shift from what was originally seen as untreatable—infertility—to something that is treatable has an impact, not just on those who now may be able to get pregnant who could not before, but also on the sense of what pregnancy is for any community aware of ART. Now the infertile woman or man is not broken permanently but possibly temporarily.

In such a sense, brokenness can appear a temporary event, but one in which one's sense of health now becomes intricately tied up with the progressive, ongoing field of medical science. In relationship to that field, one is encouraged to not just see it as a bringer of the desired capacity to have a child, but to see that one has a very particular kind of responsible relationship toward the medical world. One learns to be a very good patient.

While ART is not a universal experience, increasingly illness is always, like infertility, inserted within a progressive medical field. How I understand health as a middle-class person living in the developed world in the second decade of the twenty-first century—is radically different than others in strong dissimilar situations or at times in which the science, technologies and cultural norms around health had different forms. If I am very ill, I seek medical attention. If I am ill but the severity is unclear, I entertain the possibility of going to the doctor or taking some medication. Health and illness are always understood within a world in which medical care is integral to the understanding of health. It is hard for me to know how I would relate to illness in a world where there was no institutional medical care, where one suffered and perhaps prayed to one's god and perhaps some healers offered some therapies, but the medical *profession* as we know it now did not exist, hence no such concept existed. Critically, the existence of this field of knowledge and my awareness of it encourages thinking about health and illness as transitory states that may be altered by personal and medical practices. The broken ill body thus is always potentially curable.

I can easily imagine medical care being denied to me, or not being able to afford medical care, or medical care not being able to aid me, but in all such cases medical care still operates as the background in which I understand my health and illness. Like everyone I know, I was born in a hospital. My entire existence is pervaded by contemporary medical care just as my food consumption is determined by capitalist food production. I understand the medical profession not as a collection of people with some skills in healing, but as a complex, progressive, professional, scientific and technological set of institutions. Even if nothing is currently capable of aiding me or prolonging my life at this moment, I know that in the future such advances may happen. Such a view of the progressive nature of health care is, historically speaking, quite new. It is integral to how we understand our health and why we feel such responsibility for it since we must take up the mantle of obeying medical recommendations, seeking proper care and tracking what we can contribute to the medical field's recommendations for us. In the United States, one must also manage the complex health insurance process in order to obtain quality care.

Phenomenology of Pre-pregnancy

When one knows or is unsure of why one isn't able to become naturally pregnant, one enters into the world of testing. Like any possible disease, infertility treatments are initially tests to determine your level of brokenness and then the course of fixing. Each year since 35, I have gotten a mammogram since early breast cancer runs in my family. If I receive a result that requires more testing, obviously what this does to my sense of my future, my relationships and my perception of myself is significant. Yet, going to get tested for fertility is much different. Even though I was not initially the reason for IVF, I had to also be examined and it was discovered, not unsurprisingly, at 37-years-old that my likely supply of eggs was diminished. The doctor drew us a little diagram of how I was at the beginning of an increasingly rapid diminishing slope that would, as he pointed out very matter of fact, rapidly decline, like a stock market crash in a year or so with his pen tracing the downward slope as he articulated a series of statistics to us. Yet, to find out that having a biological child will be difficult, if not impossible, is not the same as finding out you might die of cancer. A test that indicates I might have cancer and a test that indicates I might not have a biological child produce very different implications for one's sense of brokenness.

The *possible* future being that might come into existence now became completely intertwined with this real medical practice. I counted myself fortunate to live in a subculture within the southern United States where a tight interconnected family life is front and center in every social and public situation (America's "family values"). We have many friends who live alternative paths to this view. It was not hard to imagine a life without children, even if I hoped for some, but I can only imagine how complex mourning must be for those deeply embedded in a culture where not having children would be seen a deep failure. Insofar as one might think of infertility as a "disease" and suffering from it to have features of an illness, its being about this not-yet-being make it very different than other kinds of illnesses that are about a being very much present. Any illness causes a change in one's temporality. As Merleau-Ponty famously points out, in illness one's natural reach into the world becomes stunted and the world closes in around the body.[8] One's world and one's relationships in even short-term illnesses become affected. In long-term ones, everything changes, and the future becomes radically altered and thus so do the futures of everyone one is invested in.

Yet, as phenomenologies of pregnancy point out, it is more difficult to speak about beings that are (not yet) independent of the mother. Silvia Stoeller highlights how being pregnant provides the woman with an entirely other kind of temporality—of the changes in pregnancy, the transformations, and of course the waiting for the birth, the anticipation of after pregnancy and the fears that color bad days.[9] Jonna Bornemark highlights how the a-subjectivity of pregnancy, the lack of a clear subjectivity one can assume the fetus has, points us "toward the formation of subjectivity and toward form-taking matter, or life-force."[10] One has to consider both how one's own subjectivity is itself formed out of living system that itself cannot be seen as really subjective or not. "This system is alive and thus experiencing, but the experience takes place 'everywhere.'"[11] Bornemark sees this continuing not just in pregnancy, but in many aspects of maternality—"in the maternal (pregnancy, childbirth, breast-feeding) extraordinary experiences of a-subjectivity" exist within "an already developed subjectivity."[12] She notes that such a discussion goes beyond the limits of traditional phenomenology and thus one must call for resources outside traditional phenomenology (for Bornemark, this is best considered with Deleuze and psychoanalytic research). In a similar vein that pushes beyond phenomenology, Oksala argues that pregnancy gives us "a need to rethink such fundamental phenomenological questions as the possibility of an eidetic phenomenology and the limits of egological sense-constitution."[13] The experience of thinking about this possible future being presents a different challenge to phenomenology. One is stretched toward a future where one may or may not experience this situation and to experience it one now needs technological assistance.

Testing makes the horizon of one's future shift from day to day. If one has embarked upon ART, one now will need to consider how to make one's day fit around such testing and will regard each test with concern. Will this be the end of this journey? Will this medication produce the proper result in my body? Testing shapes temporality, not just in the sense of the horizon that might hold a child or not shifts with the medical practice, but also insofar as temporality returns again and again to the past. The virtual possible child is tied up with evaluations of if one would need such elaborate measure if one's life had been different, if one had made different choices perhaps. Time extends strangely in the future even when one has received a bad result since the progressive nature of ART often means that one can always try again. Another IVF trial, or now perhaps an egg donation, a sperm donation, an embryo donation. Or perhaps there will be some new advancements that currently aren't available but might be soon enough.

Unnatural Women

One important trend is to emphasize the way in which pregnancy highlights phenomenology's androcentric tradition where simply adding more gendered analyses would fail to capture the radical differences between pregnant and non-pregnant embodiment, differences that require not just making philosophy less explicitly and implicitly sexist, but perhaps reconfiguring even what one understands philosophy to be. Luce Irigaray famously pushes philosophy to recognize that its androcentrism has missed, not just a full consideration of women, but also its main epistemological, scientific and ethical targets in its alleged "objectivity." In *An Ethics of Sexual Difference*, Irigaray writes that "characteristics" of proper epistemologies of science highlight an "isomorphism with man's sexual imaginary" and include the idea that "our personal experiences can never be used to justify any statement," but she goes on to write that "it is apparent in many ways that the subject in science is not neuter or neutral."[14] For Irigaray provocatively, overcoming androcentrism is not to create a better, more gender-neutral philosophy but rather to explore those parts of existence that refuse to be neutered.

When discussing the condition of women, Beauvoir wrote that "the man most sympathetic to women never knows her concrete situation fully."[15] Can a person who has never been pregnant understand the concrete reality of pregnant embodiment? Merleau-Ponty is often characterized, like Young does, as presenting important tools to think about tacit lived experience, but criticized for failing to consider the relevance of gendered experience. Irigaray celebrates Merleau-Ponty but argues that he did not fulfill the real significance of his phenomenology. Instead of his gender, it is his ocularcentrism that can be seen to "blind" Merleau-Ponty to "interuterine life."[16]

> My reading and my interpretation of the history of philosophy agree with Merleau-Ponty: we must go back to a moment of prediscursive experience, recommence everything, all the categories by which we understand things, the world, subject-object divisions, recommence everything and pause at the "mystery, as familiar as it is unexplained, of a light which, illuminating the rest, remains at its source in obscurity."[17]

In his lectures in child psychology and pedagogy, Merleau-Ponty does offer extensive analyses of gendered experience, including pregnancy. In the lecture "The Adult's View of the Child," Merleau-Ponty argues that pregnancy is a "major mystery" that brings the pregnant woman to "the order of life," saying, "During the entirety of her pregnancy, the woman is living a major mystery which is neither the order of matter nor the order of the mind, but, rather, *the order of life.*"[18] Following Hélène Deutsch and Simone de Beauvoir, Merleau-Ponty imagines pregnancy to bring the woman to no inconsiderable ambivalence, given her loss of a unitary and transparent selfhood. The idea of an order of life that stands in the background of the subjective perceptual experience foreshadows Merleau-Ponty's later work on flesh. While most feminist embodiment theorists that draw upon pregnant experience do not consider his Sorbonne lectures in child psychology, most do draw attention to the themes of flesh and intertwining in Merleau-Ponty's late work.

Yet, with ART, there is no mystery. One's inner body, prior to pregnancy and during pregnancy, is constantly explored and made visual. For natural pregnancies, one might not even be aware of being pregnant for an extended period of time, one may have few, if any, ultrasounds. Instead of the fetus being hidden in early pregnancy, in IVF, nothing is hidden; everything is open. Not only has one's uterus been viewed via ultrasound and tested for abnormalities prior to implantation, at the moment of implantation one is awake and aware, with a team of people surrounding you looking at exactly what is happening on a set of screens. A nurse will glide the x-ray over one's body while the doctor will discuss what she is doing—see, she is placing this embryo here and this one there. Once pregnant, I had ultrasounds every two weeks until the third month, when it became once a month. I saw my son, and his initial twin, develop from very small beans to slightly larger beans, to beans that I could understand were moving, to a thing with a head, and now little arms and legs. I laughed when the nurse proclaimed him to be a boy— saying, "Oh yes, ma'am, sure he's a boy—see!" as if this was an obvious visual. While it was clear this was a moving thing with the promise of a head and a body, it was not clear his body was gendered. From then on, we called him David, further creating him as not-me, but the future him. At the end of each session, ultrasound pictures would be printed, which we could take home. If we liked, we could also purchase videos, mugs and other products to give to family with the images of the fetus.

My son was never some indeterminate being secretly growing inside of me, but this particular fetus that I would see every two weeks. When he moved and kicked, I had images to associate with him. Thus, it was also very difficult to think of him as anything but this future being. I had no ambivalence or ambiguous sense of me-not-me described in many pregnancies, since from the moment of implantation, I had images to circulate my sense of him around. He always was the baby even when he was the fetus. Even before he was successfully lodged in my uterus in a medical setting complete with five people in medical masks, all the time spent in ART we could imagine the past and the future for this possible being we hoped to create with aid.

ART completely blends pregnancy with intensive medical care. I had no moment of pregnancy that wasn't entirely colored by high tech, at least bi-monthly medical testing. I was an excellent patient. I followed all the recommendations. I never missed a test. I did not drink or smoke or take on any stressful work. I ate well, exercised moderately, slept and took my vitamins. I quit all medications that had any possible side effects. We did not do natural childbirth because the doctor did not recommend it. We had an induced labor as well, which was recommended since we knew we could have our OB/GYN to whom we had become attached. I was very well normalized by the medical model.

Medical practice always is occurring in an existentialist frame—it is about certain, particular living individuals and their needs. "Healthiness is—as is its opposite also—a totally individually made situation" writes Hubertus Tellenbach.[19] Infertility in particular is diverse in how it impacts individuals. One reason is that while each person will experience an illness in his or her own manner, one might be able to distinguish some commonalities given how the disease typically develops into awareness and the side effects of common treatments. However, the reasons someone goes to a fertility clinic can be quite diverse. It could be to become pregnant without a male partner, it could be due to age, it could be due to earlier disease, it could be due to a condition, it could be due to the male partner's issues, which themselves are varied—low-sperm count, earlier vasectomy, disease or injury induced impotence. How the woman sees herself and her body is thus quite different whether it be her own body's infertility or that of her partner or that of not having a male partner. Whether one has some children or no children would also shape one's sense of the prognosis.

One feature of the rooms in which one waits for the blood to be drawn, the next test to be conducted was the presence of shiny flyers on various health groups, such as "Eating for Fertility," and other meet-and-greets. One dark set was directed at those who were not—at least yet—successful and needed grief counseling. One might experience great anxiety finding out that one cannot have children without medical assistance thinking one's body is not properly constituted. This kind of brokenness would mirror many other illnesses that bring the body front and center instead of allowing it to "naturally" fade into the background. Social perceptions of unnaturalness would pervade such an experience.

In addition, there is the sense of something unnatural, that is not so much about a body that all along has been diseased or disabled and can therefore be regarded as tragic, but rather about one's poor self-care. Rather, it is that the woman did not do what is properly natural at the *proper time*. Many women need ART because they wish to have children beyond the best childbearing years and their fertility is thus in rapid decline. This trend of women putting off their childbearing years for a variety of reasons (but often linked to education, professional employment and not being in "traditional" relationships) has received a great amount of editorializing by the popular presses, explicitly or implicitly suggesting that such careerist, unattached woman have failed to put family first. Even if one finds such moralizing ridiculous, the widespread knowledge of the decrease in fertility in women increasingly marks the experience of many women who think that pregnancy might be something they would wish for in life but not at this moment. This radically separates the experiences of women and men regarding pregnancy, not just for the obvious reasons of who can be pregnant, but for the common temporal limitations that men do not experience (even if their own fertility also decreases commonly with age).

Conclusion

When Irigaray accuses Merleau-Ponty of being blind to "interuterine life," she speaks of parts of the human condition that would pervade our entire history. Today, however, medicalization makes much of our self-understanding inescapabilty mediated by procedures, medicines and care-directives unknown in prior times. Peter Conrad notes that "medical designations are increasingly defining what is 'normal,' expected, and acceptable in life."[20] He goes on to discuss how impotence and sexual performance difficulties are no longer part of the human condition in particular common with age, but conditions that are expected to be treated. Likewise, ART has created a world in which women's childbearing years now have become viably extended. The future for someone who may be able to become pregnant with assistance is not just about deciding if one wishes to pursue ART, it is very often deciding at what point, if any, one should stop.

A vibrant discussion surrounds the importance of providing the capacity for persons to end their lives, with medical assistance, due to current or impending suffering due to a fatal disease. In such cases, many think that one can stop such engagement with medical assistance that can extend one's life, but perhaps shouldn't. In such a sense, one refuses to let one's broken state be temporarily solved knowing this short-term situation is just that—*short*—and its detractions outweigh its benefits. Yet, it remains much murkier to refuse to ameliorate one's brokenness when one can hope for the future of another possible person.

The power to see any kind of human condition as one that can be improved if not now, then in the future might seem to indicate the end of brokenness. Impotence, infertility, cancer and even old age all can be seen as problems to be fixed rather than something humankind must reckon with as part of our existential condition. In the *Pensées*, Blaise Pascal writes "Man is but a reed, the most feeble thing in nature; but he is a thinking reed. The entire universe need not arm itself to crush him. A vapour, a drop of water suffices to kill him."[21] I could die today of everything from someone texting while driving to poisoned food, from a brain aneurism to, as one says in the US, an "active" shooter. So many of life's absurd ways of dying are outside one's sphere of control. Yet, it is also the case that progressive medical science can help us cure what is broken in us, thus providing us with knowledges and technologies where our fragility becomes something one *can* and *should* overcome.

After contemplating the fragility of human life, Pascal calls upon the reader to contemplate the powerless of the human condition and the nobility of thought. Rather than extending into time and space, by bettering our bodies, we should rather thoughtfully reflect:

> But, if the universe were to crush him, man would still be more noble than that which killed him, because he knows that he dies and the advantage which the universe has over him; the universe knows nothing of this. All our dignity consists, then, in thought. By it we must elevate ourselves, and not by space and time which we cannot fill. Let us endeavour, then, to think well; this is the principle of morality.[22]

By contrast, now I interpret my brokenness as not something to accept about my body and then to turn toward my mind, or soul, but rather simply it is something to work toward fixing. Earlier, it was pointed out how advances in medical technologies can make what might be tragic, not simply fall into the category of disease—such as impotence or infertility. Subsequently, one is now ill instead of experiencing the wide, random and often cruel diversity of human bodies. Despite the value of contemporary cures and the likelihood of even more refined ones soon, brokenness has become in many ways more entrenched rather than cured by medical advances. No part of our embodied experience can be taken as is; no disease, no disability, no condition cannot possibly be altered to the proper "normal" or desired state. Little room exists in the human body or soul to discuss when to end one's engagement with medical practices. The progressive nature of medical science becomes mirrored in the human—any brokenness can now be addressed. If not now, perhaps in the future, or for a future person—perhaps that one a woman now carries in her uterus.

Notes

1. Talia Welsh, "The Order of Life: How Phenomenologies of Pregnancy Revise and Reject Theories of the Subject," in *Coming to Life: Philosophies of Pregnancy, Childbirth, and Mothering*, eds. Sarah LaChance Adams and Caroline Lundquist (New York: Fordham University Press, 2013), 294.
2. Nicholas Smith, "Phenomenology of Pregnancy: A Cure for Philosophy?" in *Phenomenology of Pregnancy*, eds. Joanna Bornemark and Nicholas Smith (Huddinge: Södertön University Press, 2015), 15–49, 41.
3. Iris Marion Young, "Pregnant Embodiment: Subjectivity and Alienation," in *On Female Body Experience: 'Throwing Like a Girl' and Other Essays* (Oxford: Oxford University Press, 2005), 46–61, 49.
4. Sheila Lintott, "The Sublimity of Gestating and Giving Birth: Toward a Feminist Conception of the Sublime," in *Philosophical Inquires into Pregnancy,*

Childbirth, and Mothering, eds. Sheila Lintott and Maureen Sander-Staudt (New York: Routledge, 2012), 237–50, 243.

5. Smith, "Phenomenology of Pregnancy," 41.
6. Some seminal texts in phenomenologies of illness that draw this distinction: Havi Carel, *Phenomenology of Illness* (Oxford: Oxford University Press, 2015); S. Kay Toombs, *The Meaning of Illness: A Phenomenological Account of the Different Perspectives of Physician and Patient* (Amsterdam: Kluwer Academic Publishers, 1993); Frederik Svenaeus, *The Hermeneutics of Medicine and the Phenomenology of Health* (Linköping: Springer, 2001).
7. Bjørn Hofmann and Fredrik Svenaeus, "How Medical Technologies Shape the Experience of Illness," *Life Sciences, Society and Policy* 14, no. 3 (2018): 5–6.
8. Maurice Merleau-Ponty, *Phenomenology of Perception*, trans. Donald Landes (New York: Routledge, 2012), 147.
9. Silvia Stoller, "Gender and Anonymous Temporality," in *Time in Feminist Phenomenology*, ed. Christina Schües, Dorothea Olkowski and Helen Fielding (Bloomington, IN: Indiana University Press, 2011), 80–1.
10. Jonna Bornemark, "Life Beyond Individuality: A-Subjective Experience in Pregnancy," in *Phenomenology of Pregnancy*, eds. Jonna Bornemark and Nicholas Smith (Huddinge: Södertön University Press, 2015), 276.
11. Ibid.
12. Ibid.
13. Joanna Oksala, "What Is Feminist Phenomenology? Thinking birth philosophically," *Radical Philosophy* 26 (July/August 2004): 17.
14. Luce Irigaray, *An Ethics of Sexual Difference*, trans. Carolyn Burke and Gillian C. Gill (Ithaca, NY: Cornell University Press, 1993), 122.
15. Simone de Beauvoir, *The Second Sex*, trans. Constance Border and Sheila Malovany-Chevallier (New York: Knopf, 2010), 14–15.
16. Luce Irigaray, *An Ethics of Sexual Difference*, 152.
17. Ibid., 151.
18. Maurice Merleau-Ponty, *Child Psychology and Pedagogy: The Sorbonne Lectures 1949–1952*, trans. Talia Welsh (Evanston, IL: Northwestern University Press, 2010), 101.
19. Hubertus Tellenbach, "The Phenomenology of States of Health and Its Consequences for the Physician," *Morality within the Life—and Social World* 22 (1987): 374, doi: 10.1007/978-94-009-3773-4_27
20. Peter Conrad, *The Medicalization of Society: On the Transformation of Human Conditions into Treatable Disorders* (Baltimore: The Johns Hopkins University Press), 149.
21. Blaise Pascal, *Pensées*, #347. Project Gutenberg, www.gutenberg.org/files/18269/18269-h/18269-h.htm
22. Ibid.

13 Dying Bodies and Dead Bodies

A Phenomenological Analysis of Dementia, Coma and Brain Death

Fredrik Svenaeus

Introduction

How should phenomenologists of medicine and bioethics approach death? Who or what is dying when a patient is terminally ill: the person or his/her body? In most cases when a person dies these two things happen more or less simultaneously—the person ceases to exist precisely because his/her body fatally breaks down—but there are some cases when the life and death of a person and his/her body split ways. The main examples of this split in the present chapter will be late stage dementia, coma, and brain death in which the body appears to have survived the life of the person. The opposite science fiction scenario will not be dealt with here, but it should be noted that the transhumanist dream of an eternal (or very long) posthuman life is precisely the dream of a person surviving his/her current body by uploading the neuronal patterns of the brain into a computer.[1] Such thought experiments about what may be possible in a future world of brain scanners and tele-transporters have been very influential in philosophy of mind and moral philosophy for a long time.[2] What is most often neglected in such analyses is precisely embodiment as viewed from the phenomenological first-person perspective. The life of a person is not the life of a mind, or even of a brain, but the life of an experienced and experiencing body. As Maurice Merleau-Ponty puts it in *Phenomenology of Perception*:

> The body is the vehicle of being in the world and, for a living being, having a body means being united with a definitive milieu, merging with certain projects, and being perceptually engaged therein. . . . For if it is true that I am conscious of my body through the world and if my body is the unperceived term at the center of the world toward which every object turns it face, then it is true for the same reason that my body is the pivot of the world.[3]

And further concerning the dying and dead body:

> Science accustoms us to considering the body as an assemblage of parts, and so too does the experience of its breaking apart in death.

Now, the decomposed body is precisely no longer a body. If I put my ears, my nails, and my lungs back into my living body, they will no longer appear as contingent details. They are not indifferent to the idea of me that others form, they contribute to my physiognomy or to my style. . . . In other words, as we have shown elsewhere, the objective body is not the truth of the phenomenal body, that is, the truth of the body such as we experience it.[4]

My strategy in this chapter will be to turn to the notorious analysis of "being-toward-death" found in Martin Heidegger's *Being and Time*, supplemented by some interspersed reflections from other phenomenologists on death and dying.[5] However, in order to prepare the reader to Heidegger's terminology and the phenomenological points he is making, I will first say some things about the notion of personhood as related to the life of early human beings. This is a good way to introduce issues about the *gradual* appearance of personhood in the case of human beings that are important for the arguments I want to make regarding the gradual *dis*appearance of personhood in other cases. After discussing Heidegger's views on death, I will then turn to the examples of dementia, coma and brain death and the way these have been interpreted in bioethics as concerns the late life of persons and their bodies. Along the way, I will, in addition to discussing the phenomenology of different stages of personhood, also make some points about how such an ontology provides us with moral insights. In connection with identifying some key positions in bioethics regarding the definition of life and death of human beings, I will finally state the way I think we should conceive of dying and death in the discussed examples from the phenomenological point of view.

The Phenomenology of Early Personhood

Persons are standardly defined in contemporary philosophy as creatures possessing self-consciousness, language, memory and an ability to plan their actions.[6] In *Being and Time* Heidegger employs the terminology of "Da-sein" when he refers to such person-shaping characteristics. To have Da-sein roughly means to be aware of oneself as a subject that is placed in a meaningful environment together with others, what is also referred to as a "being-in-the-world."

As many philosophers have pointed out, Heidegger is remarkably silent on the meaning of birth in comparison with death in *Being and Time*.[7] I think the most tenable interpretation of his view on early human life is that children *gradually* come to inhabit a being-in-the-world when they establish understanding relationships to other subjects and objects that address them and surround them through their infancy.[8] However, even before being born a child becomes *sentient* in the womb and this is surely also a relevant occurrence from the perspective of personhood. As Dan

Zahavi and others have argued, the most basic form of selfhood consists in the *feeling* of having experiences in the first place.[9] Let us call the fetus, which becomes sentient in the womb around week 24 gestational time,[10] a *very early person*. The fetus has embodied experiences of being alive and feeling its various states to be pleasurable or painful even though these experiences are certainly not intentional in the sense of *being about* other things inside or outside the mother's body.

The next significant stage in the life of a fetus is the birth that makes it into a baby. This is the event that phenomenologist Hans Jonas associates with the ethical appeal to shoulder responsibility for the child's vulnerable and dependent being.[11] Birth is significant from an ethical point of view because the baby *presents* itself to the world, that is, to the persons who are there to take care of it, and this ushers in a different kind of responsibility than the fetus in the womb is capable of appealing for. Through birth the baby becomes, to use a metaphor found in the work of Emmanuel Levinas, a very early person with a *face*.[12] To have a face in this context does not only mean to have the physical characteristics of eyes, nose and mouth in place, it means that the child expresses a vulnerable, personal being by way of how it looks, sounds, feels, smells and so on.

However, from the point of view of development of personhood, physical birth is not the next decisive event in the life of a child after becoming sentient. This is, rather, the event that is sometimes called "psychological birth": the child's opening up to the world around it, and, most significantly, the communicative expressions and responses offered in face-to-face interaction with the parents (or other care persons).[13] This is called the "two-month revolution" in the development of the baby, and it shows that around this age the baby is not only experiencing basic feelings but is also entering the shared, intersubjective realm of a being-in-the-world in a lived bodily way (compare the quotes from Merleau-Ponty above).[14] Let us call the six-week-old baby an *early person*, who is now sharing the world with others and through successive developmental stages will become aware of things and persons in the world around him/her, and, also, of him/herself as an embodied creature capable of acting and expressing his/her wishes.

The next decisive step—I am certainly skipping most of the details in this sketchy overview—from the perspective of personhood is the one associated with passing what is known as the "mirror test"[15] This usually commences around 18 months of age and is supposed to prove that the child—or animal other than *Homo sapiens*—is aware of him/herself in an explicit *reflective* way.[16] The test consists in putting the child in front of a mirror, but first, without the child's awareness, marking its forehead with a clearly visible dot or sticker. Before the age of 18 months (roughly) the child will point toward the dot or sticker seen in the mirror; after 18 months it will put its hand on its own forehead, realizing that the child in the mirror reflection is identical with itself.

Philippe Rochat, the well-known child developmental psychologist to whose path-breaking work, *Others in Mind: Social Origins of Self-Consciousness*, I have referred earlier, convincingly argues that the birth of self-consciousness in a child is not only a cognitive move but also, and more importantly, an emotional recognition of oneself as somebody being seen and evaluated *by others*. Reflective self-consciousness means that the child becomes aware of itself as a "me" in the eyes of others in addition to a pre-reflectively experiencing "I," and this first feeling of me-ness is most often a feeling of embarrassment or shame.[17] Interestingly, other animals than humans that have been reported to pass the mirror test (primates, corvids, dolphins, elephants) do not as a rule express such social feelings (at least not in a way that we humans fully understand). Complex social emotions appear to be unique to humans (and possibly some other species of the great apes), and they show the extent to which we as persons are dependent upon a network of social relations that demand a moral sensitivity nurtured by these very feelings.[18]

From the age of one and a half years the child is thus a *person*—having passed the very early and early stages—being aware of itself in a world shared with other persons. This standardly includes some use of language and a rudimentary memory as well as the capacity to plan actions ahead, so we have all the characteristics usually associated with personhood in place.[19] However, the understanding of social roles and moral obligations are still very primitive and limited in the second year of child's life. Over the next three years the language abilities and temporal understanding of the child will develop in substantial ways. This allows the emergence at around four and a half of what we could call a *narrative person*, with a picture and story of him/herself in which judgments concerning moral obligations to others and personal ideals have been formed.[20]

When Heidegger refers to the life of "Da-sein" and the understanding it can establish as concerns its own way of existing, he is most often referring to what we in terminology developed above would call narrative persons. As we will see, in analyzing the processes of dying it becomes relevant to also address more rudimentary forms of personhood corresponding to the full but non-narrative, early and very early forms of personhood we have surveyed and defined previously. We will return to the different levels of personhood. Let us now take a look on how Heidegger views death issues in the framework of Da-sein and its being-in-the-world.

Heidegger on Death

> *No one can take the other's dying away from him.* Someone can go "to his death for an other." However, that always means to sacrifice one-self for the other "*in a definite matter*". Such dying for . . . can never, however, mean that the other has thus had his death in the least taken away. Every Da-sein must itself actually take dying upon itself. Insofar

as it "is," death is always essentially my own. And it indeed signifies a peculiar possibility of being in which it is absolutely a matter of the being of my own Da-sein. In dying it becomes evident that death is ontologically constituted by mineness and existence. Dying is not an event, but a phenomenon to be understood existentially in an eminent sense still to be delineated more closely.[21]

Heidegger claims that in a phenomenological (other terms he is using are "ontological" and "existential") analysis death (*der Tod*) should not be understood as an event but as a way of understanding one's own way of being. The German verb he uses to express the personal understanding of finitude as belonging to my own way of being is "*sterben.*" Dying is in this sense something that belongs to Da-sein's (a person's) way of *existing* as such. We are dying continuously at every moment from the point of understanding that we are finite creatures (probably we start doing so somewhere in between becoming persons and full narrative persons on the timeline developed above). However, Heidegger claims that this is an insight we are constantly shying away from because it gives rise to anxiety. Instead of facing the anxiety nurtured by the insight that every moment could be our last and live according to this basic truth, we tranquilize ourselves by pretending that death is something that happens only to others:

> The publicness of everyday being-with-one-another "knows" death as a constantly occurring event, as a "case of death". Someone or another "dies," be it a neighbour or a stranger. People unknown to us "die" daily and hourly. "Death" is encountered as a familiar event occurring within the world. . . . The public interpretation of Da-sein says that "one dies" because in this way everybody can convince him/herself that in no case is it I myself, for this one is *no one.*[22]

Heidegger is employing quotation marks to indicate that the common way of understanding death and dying is not the phenomenological way but the way of whom he refers to as "the one" (*das Man*). In an analogous manner to the medical view on death, public discourse reduces dying to a material event, something that happens to other people and maybe also to oneself in the future, but certainly not right now. In contrast to this, dying in the qualified phenomenological sense is not a physiological, or even a psychological, process but an *attitude* to life in which the anxiety of finitude is accepted and developed into a positive mode of understanding your own possibilities.[23] In this way dying, in the phenomenological sense, is an anticipatory being *toward* the end, which needs to be constantly affirmed and developed by a person as long as she lives.

Heidegger aims to develop dying (*Sterben*) into a key existential concept loaded with ethical significance. He refers to the phenomenological understanding of death as "authentic" in contrast to the inauthentic understanding of death employed in everyday discourse and medical science. And he develops this understanding as preconditioned by accepting anxiety in the "face of death," listening to "the call of conscience" summoning Da-sein to its own "resoluteness."[24] In the end "being-toward-death" will prove to be a way of understanding the being of Da-sein from a temporal perspective, as indicated by the title of the book *Being and Time* itself.[25]

It is well beyond the purpose of this chapter to further disclose and discuss the merits and shortcomings of Heidegger's ethics of authenticity. In any case, such a temporal-existential analysis will not discharge us of a phenomenological analysis of the particular circumstances in which we come close to our own death (or the death of close others) because of illness or other fatal happenings. Such happenings may open up extraordinary possibilities for authentic being-toward-death in Heidegger's sense, but they also make obvious that we need other concepts than being-toward-death to make sense of the things that happen when people die.[26] As a matter of fact, Heidegger himself attempts to develop such a terminology in his book. After discussing scientific investigations of the life and death of plants and animals—"ontic" investigations in contrast to the phenomenological "ontological" investigations—he writes:

> An ontological problematic underlies this biological and ontic investigation of death. We must still ask how the essence of death is defined in terms of the essence of life. The ontic inquiry into death has always already decided about this. More or less clarified preconceptions of life and death are operative in it. These preliminary concepts need to be sketched out in the ontology of Da-sein. Within the ontology of Da-sein, which has priority over an ontology of life, the existential analytic of death is subordinate to the fundamental constitution of Da-sein. We called the ending of what is alive "perishing" (*Verenden*). Da-sein, too, "has" its physiological death of the kind appropriate to anything that lives and has it not ontically in isolation, but as also determined by its primordial kind of being. Da-sein, too, can end without authentically dying, though on the other hand, qua Da-sein, it does not simply perish. We call this intermediate phenomenon its "demise" (*Ableben*). Let the term "dying" (Sterben) stand for the *way of being* in which Da-sein *is toward* its death. Thus we can say that Da-sein never perishes. Da-sein can only demise as long as it dies.[27]

This admittedly rather long quote from *Being and Time* contains all the distinctions we need to make sense of death in the context of human beings, also in cases when the life of persons and their bodies split ways, as I put it in the introduction earlier. Organisms (plants, animals, the human animal) merely "perish" when they die in the terminology of Heidegger. Persons never perish, they "demise" when they go out of existence and they constantly "die," since this is the pattern of their characteristic way of self- and world-understanding. Heidegger's terminology is, indeed, nonstandard when compared to contemporary everyday ways of talking about death, and, also, as concerns the terminology of bioethics. My final verdict will be that we should use the terms death and dying in other ways than Heidegger does, but that the distinctions he develops between the ways bodies and persons live and die are fundamentally right on spot. In order to show this, I will now turn to a closer examination of, first, dementia, and, then, coma and brain death.

The Phenomenology of Late Personhood

When we become old our bodies inevitably display the kind of human vulnerability we have been suffering from ever since we were born.[28] Transhumanists dream of an age when we will no longer have to die, because doctors and other scientists will be able to fix or replace our aging body.[29] However, for a foreseeable future we will have to live with our vulnerable condition, which means that aging inevitably brings maladies and illnesses of different sorts and the suffering that follows in their tracks. We adapt to this increasingly vulnerable and weak condition with the more or less spontaneous change of lifestyle that often commences in growing old. Old people live slower and more cautious lives; they become less focused upon doing new things and treasure relationships with people they already know.[30] To become older means embodying a life narrative that is coming to a close, and this is not necessarily a bad or sad thing.

In using the expression "embodying a life narrative," I literally mean a person's lived embodiment as the central aspect and way of existing in a world with others. Ways of embodiment change with age, and this is also the reason persons modulate or change the preferred life projects from which they derive their core life-narrative values.[31] We generally become less physically active and more thoughtful as we grow older; we often care less about our shortcomings and appreciate the things we are still able to do.[32] In this manner we can escape alienation and even become more at home with ourselves in getting closer to the end of our life.

Analogously to the way persons gradually come into being as first early and then full narrative creatures they can also, in some cases, gradually disappear. In Heidegger's terminology we would say that these persons *demise* by suffering a kind of slow departure from life. Alzheimer's is a chronic neurodegenerative disease that usually starts slowly and gets worse over time. It is the major cause of dementia, which may also be caused by other diseases and injuries that affect the brain, such as stroke.[33] In the early stage of dementia, the person suffers from symptoms such as forgetfulness, losing track of time and getting lost in familiar places. In the middle stage, the person may become forgetful of recent events and people's names, get lost in his/her home environment, have difficulties finding the right words, need help with personal care and go through behavioral changes. In the late stage, the person becomes unaware of time and place, has difficulties recognizing family members and close friends, becomes unable to take care of basic needs such as eating, dressing and going to the toilet, loses the ability to walk and may suffer personality change. Alzheimer's disease and most other forms of dementia lead to death. In the last stage, the patients are most often bedridden, conscious and able to respond to verbal or non-verbal address, but are totally bereft of cognitive understanding of who they are and what is going on around them.

Dying from Alzheimer's, or dementia by another cause, is a cruel death, not least for relatives and close friends of the patient, but it is also an interesting process for phenomenological studies of personhood. It appears that in the case of going through the different stages of gradually progressing dementia we have a kind of mirror image of the gradual appearance of a person in infancy and childhood. The early and middle stages of dementia correspond to the gradual appearance of (in the case of dementia, the gradual disappearance of) the narrative person. In the late and last stages, we witness the person deteriorating gradually toward the form of early personhood that comes into being about six weeks after birth (the psychological birth of the baby) or even the very early forms preceding this event in which the baby or fetus is conscious and able to feel good or bad but nothing much beyond this.

The stages that the Alzheimer's patient goes through in suffering and dying from this terrible disease we could call the stages of a *person* with a gradually deteriorating narrative, a *late person* and, in some cases, a *very late person*, corresponding to person (18 months after birth), early person (six weeks after birth) and very early person (about week 24 gestational time).[34] Beyond this we have the stage of coma that would be the equivalent to what we could name the life of a non-sentient *pre-person* in the womb. Let us call such permanently comatose patients *post-persons*.

Do late persons have the same moral status as early persons, very late persons the same moral status as very early persons and post-persons the same moral status as pre-persons? Should we, in each case, treat them in the same way with regard to dignity, protection-worthiness, human rights and the like? No, this does not follow, since we are in the cases of persons affected by dementia in gradually progressing stages and post-persons dealing with beings who are completing rather than beginning, their life narratives. The manner in which we relate to these human beings and treat them must reflect this temporal and narrative structure that is characteristic of a human life when thought about and reflected upon.[35] We treasure a two-month-old baby, not only for who he/she already is, but also for who he/she will *become* together with us. We love and respect a severely demented person, not only for who he/she is, but also for who he/she has *been*. Yet there are some lessons to be learned from the comparisons. Post-persons are no longer persons, but they are still *living human beings*, just as the embryo and the early fetus are.

Brain Death

Is it possible for a patient to be dead while still breathing? This question arose in the 1960s and 1970s when new medical technologies made it possible to keep the heartbeat and respiratory function of brain-damaged individuals intact by connecting them to ventilators when they could no longer breathe by themselves. The reasons for challenging the aliveness of these patients were the fact that they would never become conscious again and, especially, that parts of their bodies could be used to help others in need if they were declared dead while blood was still perfusing their organs. These concerns led to the implementation of a complementary way to make a medical judgment of death in most countries of the world during the 1970s, 1980s, and 1990s: brain death. Accordingly, in addition to the traditional criterion of loss of cardiopulmonary function, a person may also be defined as dead if he/she has irreversibly lost all functionality of the brain even if the circulatory system of the body is being maintained by machines.[36]

Brain death is not the same thing as coma. Persons who are temporally or permanently comatose may still have intact brain stem functionality necessary for cardiopulmonary and other vital bodily functions. And the cerebral functions necessary for consciousness, which are absent in coma, may be only temporarily gone, as it is when people are anaesthetized, for instance. Brain death is also different from what is called a permanent vegetative state, in which a person is awake (or asleep) but not aware of what is going on. People in a coma or a vegetative state have not lost all the functions of the brain, and even though the doctors, after studying the damage done to their brains and assessing their long-term condition, may establish with very high likelihood that they will not regain consciousness, it is impossible to establish this beyond *all* doubt. These patients are kept alive by feeding tubes and nursing care and they can live for years or even decades if nutrition and care of the body is maintained. The wishes of and conflicts between relatives and medical personnel concerning whether or how to allow these patients to die are legendary in bioethics.[37]

In addition to brain death, coma and vegetative state, two related conditions should be mentioned in which consciousness is *not* totally lost: minimally conscious state and locked-in syndrome. A person in a minimally conscious state may look much like a patient in a vegetative state except that awareness can be proved beyond reflexes and automated behaviors like swallowing or blinking when he/she is exposed to external stimuli. A person in a minimally conscious state is able to understand and respond to simple questions, expressing feelings by means of body language or moving a limb when asked to do so, for instance. A person going through the late stages of dementia (a late person) may eventually find him/herself in a minimally conscious state. A person in a locked-in syndrome is fully aware and conscious (a full narrative person) despite suffering from total, or nearly total, bodily paralysis. Cognitive functions of the higher brain are intact, while damage to the lower parts of the brain prevents the person from voluntarily moving any part of his/her body with the exception, in most cases, of vertical movement of the eyes and blinking.

A person in a minimally conscious state may easily be mistaken for a patient in a vegetative state, or vice versa, because of the difficulties in establishing whether bodily responses are conscious or automated. And locked-in conscious states may easily go undetected, especially if the paralysis also affects eye movements and blinking. Locked-in syndrome appears to be not only a truly nightmarish condition but also a case of at least minimal, if severely restricted, embodiment. A locked-in person is, indeed, never totally locked in because he/she is able to see and hear what is going on around him/her and is also able to express him/herself by way of his/her eyes. Proprioception and bodily perception are also often present to some degree despite the paralysis.[38]

Brain death, defined as the irreversible loss of all functions of the brain, can be established beyond all reasonable doubt by examining the type of damage done to the brain and establishing the absence of all electro-chemical activity. In cases of brain death, the present ethical dilemma is not about whether but *when* to turn off the life-sustaining technology. The ventilator is actually not a *life*-saving technology anymore if the patient is declared brain dead, since he/she is then *dead* according to the law in most countries of the world at the present time. The ethical conflicts are instead about whether such patients should be kept in a state in which their organs are perfused with oxygenated and nutritious blood for the sake of others. Should they be treated for somebody else's—the organ receiver's—sake for a while rather than for their own good? Or, as phenomenologist Hans Jonas argued in a paper published in 1968 and continued to stress in following the implementation of brain death in medical practice and laws of various states in the United States, is this procedure nothing but the instrumentalization of the body of a patient who is still alive?[39]

It should be pointed out that Jonas' wish and ethical claim was not that permanently comatose or brain dead patients should be kept alive when the chances of their regaining consciousness and a life worth living were (close to) zero. His claim was that we are not entitled to declare them dead—rather than letting them die—in order to be able to use their bodies as "organ banks," as he puts it.[40] Jonas is not alone in criticizing the idea and definition of brain death for being incoherent and pragmatically rather than scientifically motivated.[41] There is no doubt something strange about claiming that patients with non-functioning brains who are kept on life (!) sustaining technology, and who in some cases have been witnessed to go through puberty, heal wounds, fight off diseases and even gestate a baby, are dead. These bodies, indeed, appear to have survived the death of their brains.

The main problem if we want to be able to use organs from patients with non-functioning brains for transplants is that we would *kill* them by doing so if we have not first declared them dead (the dead donor rule). An interesting possibility, which Jonas raises in the last postscript to his 1968 paper, written in 1985 when the legislation on brain death had already been put in place, is at least to turn the ventilator off and wait until the heart has stopped beating *before* removing the organs.[42] This would not be totally ideal from the perspective of preserving the organs, but the advantage would be that the medical staff would allow the patient to die—according to the traditional definition of death—before opening his/ her body.

What We Are and How We Die

The legal implementation of the concept of brain death reflects the practical concern that what I have called the post-person is, indeed, no longer a person. Permanently comatose patients are also post-persons, the main differences distinguishing these patients from the brain dead being the continuation of brain stem functionality for breathing and the difficulties of establishing beyond all doubt that they will never regain consciousness again.

David DeGrazia, in his impressive study *Human Identity and Bioethics*, argues that we are essentially human *animals*, not persons.[43] We cannot essentially be persons, because some human animals are clearly not persons (embryos and permanently comatose patients), and some persons are not human (other animals, possibly even artificially intelligent computers). I think this is a too hastily drawn conclusion that fails to take into account the significance of two different perspectives on human beings: the first-person perspective (including the second-person perspective) and the third- (or rather non-) person perspective. From the third-person perspective of science, all living human beings are biological organisms (animals). Some such organisms reach a stage of complexity and sophistication that allows for the first-person, experiential perspective to occur. The person that develops from the very early and early stages of personhood toward the stage of having a full narrative does so by way of being introduced to a shared world in which the second-person perspective is primary and in which the third-person perspective has historically arisen as a way of scientifically engaging in explanations of why things in the world work as they do (including human organisms).

I thus propose a view that is similar to the one put forward by Lynne Rudder Baker in her study *Persons and Bodies: A Constitution View*: persons are constituted (meaning: made possible) by their biological organisms.[44] This means that persons neither die nor are born in the physical sense; instead, they come to and go out of existence more or less gradually depending on the states of the biological organisms that constitute their beings. Biological organisms are internally organized things that resist entropy and reproduce, using energy from their environment.[45] This means that individual human animals—which are one species of such creatures—come to life when the embryo is created (minus a couple of weeks to make sure that no split or fusion into or with other embryos occur),[46] and die when the organism has broken down (is no longer internally organized and resisting entropy). Baker holds that only creatures that have achieved the level of what I have called narrative persons are persons, whereas I have given priority to an experiential first-person perspective without denying the significant steps taken when very early persons turn into early persons, persons, and narrative persons, respectively (or when narrative persons turn into persons, late persons, and very late persons, respectively).[47]

The relationship between a living human body and the person that the living body, makes possible under certain circumstances is similar to the relationship between disease and illness. Biological organisms become diseased but only persons are ill and suffering by experiencing their life as a being-in-the-world.[48] When we say that an organism, say a worm or a brain dead human being, *suffers* from a disease, that is, from a biological dysfunction, we are using this word in a metaphorical sense. And when we say that a person has *died*, what we really mean is that the biological organism that once constituted his/her existence is no longer capable of doing so. The proposal I am presenting in this chapter is in line with the view held by Jonas regarding the life and death of bodies, but I try to do be more specific and informative in regard of how biological processes may give rise to lived experiences had by persons in different stages of appearance (early persons) and disappearance (late persons).[49]

My interpretation is also perfectly in line with the distinctions made by Heidegger in *Being and Time* between perishing, demising and dying, save that, in my terminology, dying is reserved for that which Heidegger calls perishing instead of referring to the whole self-understanding existence of Da-sein. As Heidegger writes, life and living are not really apt words for describing the being of Da-sein, that is, the essence of personhood as I have interpreted it. Da-sein (the person) rather *exists*, whereas the biological body is the entity that qualifies as living.[50] This is consistent with the idea that the body may go on living also in cases when the person, who it once constituted, does no longer exist. Yet, there is also a different way to develop Heidegger's terminology of death in contemporary medicine and bioethics, namely to reserve the term "dying" for the processes that persons go through (both demising and dying in Heidegger's terminology) and insist that plants and animals (including the human animal) do not die in the true sense of the word, they rather perish. This is a view developed by some philosophers of death and dying in contemporary bioethics, if not supported by arguments from Heidegger or other phenomenologists, notably John Lizza in *Persons, Humanity, and the Definition of Death.*[51]

Lizza claims that what the legislators of brain death are attempting to express by way of stressing the crucial functions of the brain for the life of human bodies is that the *person* is dead, even though his body minus the brain is kept alive by medical machinery.[52] Lizza supports the constitutive view that I adopted previously, but according to him it is the *person* made possible by way of the biological organism that dies, not the organism itself.[53] Perhaps the difference between the view of Lizza and the view I have developed in this chapter is mainly a quarrel about terminology. I certainly agree with him that what the doctors are trying to say about brain dead patients when calling them dead is exactly that the person is gone. This is also what they are trying to say in cases of permanent coma or vegetative states when motivated to shut down life support or not treat diseases. There is no point in keeping a patient alive if the person that his/her living body once made possible is irreversibly gone.

Life and death have attained specialized meanings in biology and medicine based on historical, everyday uses of these concepts. As many have pointed out, life and death are notoriously hard to define due to many borderline situations. Nevertheless, we have a basic sense of what we mean by something being alive and this meaning, ever since Aristotle, refers to what we may call self-moving bodies. As I remarked earlier, organisms are currently thought about in biology as internally organized things that resist entropy and reproduce, using energy from their environment.[54] When these processes permanently stop, the organism is dead (whether bacteria are dead or only hibernating may be even harder to establish than whether a patient in a coma will recover). I think we do best to reserve the terms life and death for the state of the organism (animal) in medicine and bioethics. It is quite clear, and Lizza acknowledges this, that the notion of brain death as formulated in the law explicitly applies to human beings as biological organisms and *not* as persons.[55]

To systematically change the use of the terms death and dying to apply to persons only, or, possibly, to persons *and* organisms, would be potentially confusing and even harder to accept for the public than the application of the brain death criterion has already been. Taking into account that the concept of personhood is generally not being addressed in the embodied and continual way I have attempted in this chapter, but rather in a disembodied and reductionist manner, it would also be a risky strategy. Not only the brain dead and the permanently comatose, but also the vegetative and minimally conscious, plus some other cases of late demented persons would probably find themselves dead if personhood referred to creatures that embody what I have called full narrative personhood, only. Even with a narrower definition, corresponding to the one I have termed full, but not full narrative, personhood, the class of human beings who are dead, and whose bodies should therefore be allowed, or even made, to "perish" would raise significantly. Though this may be a good thing in consideration of the biological resources made available for living persons (organ transplantation) it would also open the gates for exploitation of human bodies in massive scale.[56]

The consequence of the phenomenological view on death that I have developed in this chapter is that neither the brain dead nor the permanently comatose patient is dead. But they are not persons, not even late or very late persons, anymore, since they are permanently non-conscious, lacking the *experiences* that are necessary for the first-person perspective to subsist. What the philosophers, doctors and politicians implementing the concept of brain death are trying to say is that the *person* is gone if the biological organism has reached a stage in which the brain has ceased to function entirely. This is right, but from the third-person, scientific perspective in which questions about *death* ought to be settled, they are all wrong: the living body that once constituted the persistence of a person is not dead; it is, rather, fatally damaged and being kept *alive* with the help of medical technology.[57]

Conclusive Thoughts about the Death of Human Beings

When the human organism that supports a person has died and left behind a corpse, the person is clearly no longer there. Yet the persons who are left behind are still *with* the deceased, and they treat the dead body as something which is connected to the person it once constituted: they grieve over it, bury it (in some way) and erect a stone (or something similar) to keep the person in their memories (and in the memories of other people still to come):

> The "deceased," as distinct from the dead body has been torn away from "those remaining behind," and is the object of "being taken care of" in funeral rites, the burial, and the cult of graves. . . . In such being-with the dead, the deceased *himself* is no longer factically "there." However, being-with always means being-with-one-another in the same world. The deceased has abandoned our "*world*" and left it behind. It is *in terms of this world* that those remaining can still *be with him.*[58]

What about post-persons in this regard? Are we with the permanently comatose and brain dead in a more substantial, different sense than is the case with corpses? In many cases we feel that we are, because their still living bodies appear to *express* the presence of a person in a way a corpse does not usually do. The presence of the permanently comatose (and possibly of some corpses, too) is *uncanny* to us in this regard, as their living bodies appear to be *lived* by a person who is gone.[59] But they are not, really, having experiences in such a way, and to the extent that doctors are able to judge with absolute confidence that these living bodies will never regain consciousness again, we do best in viewing the post-person as an anticipatory corpse.

Could we be *with* such bodies in a way that allows us to treat them with dignity: honor them, grieve for them and still transplant their organs once we have turned off the ventilator and other medical equipment that keeps them alive? I believe this is possible, but to avoid instrumentalization, the treatment of such bodies must be governed by a respect for the persons and life narratives that they previously made possible. If we insist on giving priority to a medical-scientific or liberal-economic perspective on them and neglect to honor the shared being-in-the-world that they are still related to, then commodification lurks around the corner.

Notes

1. Mark O'Connell, *To Be a Machine: Adventures Among Cyborgs, Utopians, Hackers, and the Futurists Solving the Modest Problem of Death* (New York: Doubleday, 2017), 42 ff.
2. The work by Derek Parfit, *Reasons and Persons* (Oxford: Oxford University Press, 1984), has been seminal in this regard.
3. Maurice Merleau-Ponty, *Phenomenology of Perception*, trans. Donald A. Landes (London: Routledge, 2012), 84.
4. Ibid., 455–6.
5. Martin Heidegger, *Being and Time*, trans. Joan Stambaugh (Albany: State University of New York Press, 1996). (Page references are to the German original found in the margins of the English translation).
6. David DeGrazia, *Human Identity and Bioethics* (Cambridge: Cambridge University Press, 2005), 3–7.
7. See, however, Heidegger, *Being and Time*, 373–4.
8. Regarding the phenomenology of early childhood, see Philippe Rochat, *Others in Mind: Social Origins of Self-Consciousness* (Cambridge: Cambridge University Press, 2009).
9. Dan Zahavi, *Subjectivity and Selfhood: Investigating the First-Person Perspective* (Cambridge, MA: The MIT Press, 2005), chapter 5.
10. Carlo Valerio Bellieni, "Pain Assessment in Human Fetus and Infants," *The AAPS Journal* 14, no. 3 (2012): 456–61.
11. Hans Jonas, *The Imperative of Responsibility: In Search for an Ethics for the Technological Age* (Chicago: Chicago University Press, 1984), 131.
12. Emmanuel Levinas, *Totality and Infinity*, trans. Alphonso Lingis (Dordrecht: Kluwer Academic Publishers, 1991).
13. Rochat, *Others in Mind*, 69.
14. Ibid., 67–78.
15. Zahavi, *Subjectivity and Selfhood*, chapter 7.
16. Rochat, *Others in Mind*, chapter 5.
17. Ibid., chapter 6.

18. Regarding these matters, see the works by Frans de Waal, *Primates and Philosophers: How Morality Evolved* (Princeton, NJ: Princeton University Press, 2006); and Anthony J. Steinbock, *Moral Emotions: Reclaiming the Evidence of the Heart* (Evanston, IL: Northwestern University Press, 2014).
19. DeGrazia, *Human Identity and Bioethics*, 3–7.
20. Rochat, *Others in Mind*, chapter 9.
21. Heidegger, *Being and Time*, 240.
22. Ibid., 252–3.
23. For an excellent overview and interpretation of Heidegger's concepts of death and dying, see Kevin Aho, "Heidegger, Ontological Death, and the Healing Professions," *Medicine, Health Care and Philosophy* 19, no. 1 (2016): 55–63.
24. Heidegger, *Being and Time*, 267 ff.
25. Ibid., 301 ff.
26. See Aho, "Heidegger, Ontological Death, and the Healing Professions."
27. Heidegger, *Being and Time*, 246–7.
28. Alasdair MacIntyre, *Dependent Rational Animals: Why Human Beings Need the Virtues* (Chicago: Open Court, 2001).
29. O'Connell, *To Be a Machine*.
30. Atul Gawande, *Being Mortal: Illness, Medicine and What Matters in the End* (London: Profile Books, 2014).
31. Fredrik Svenaeus, *Phenomenological Bioethics: Medical Technologies, Human Suffering, and the Meaning of Being Alive* (London: Routledge, 2017), chapter 2.
32. Gawande, *Being Mortal*.
33. Regarding Alzheimer's disease and dementia, see World Health Organization (WHO). *Dementia: Fact Sheet*, 2016, www.who.int/mediacentre/factsheets/fs362/en/
34. Regarding matters of personhood in dementia, see the essays in Julian C. Hughes, Stehpen J. Louw and Steven R. Sabat, eds., *Dementia: Mind, Meaning and the Person* (Oxford: Oxford University Press, 2006).
35. Paul Ricoeur, *Oneself as Another*, trans. Kathleen Blamey (Chicago: University of Chicago Press, 1992), chapter 6.
36. Stuart J. Younger, "The Definition of Death," in *The Oxford Handbook of Bioethics*, ed. Bonny Steinbock (Oxford: Oxford University Press, 2007), 285–303.
37. Jeff McMahan, "Death, Brain Death, and Persistent Vegetative State," in *A Companion to Bioethics*, ed. Helga Kushe and Peter Singer (London: Wiley-Blackwell, 2009), 286–98.
38. See the fascinating account offered by Jean-Dominique Bauby, *The Diving Bell and the Butterfly: A Memoir of Life in Death*, trans. Jeremy Leggatt (New York: Vintage Books, 1998).
39. Hans Jonas, *Technik, Medizin und Ethik* (Frankfurt am Main: Suhrkamp Verlag, 1987), chapter 10.
40. Ibid., 219.
41. See DeGrazia, *Human Identity and Bioethics*, chapter 4; and Younger, "The Definition of Death."
42. Jonas, *Technik, Medizin und Ethik*, 239.
43. DeGrazia, *Human Identity and Bioethics*, chapter 2.
44. Lynne Rudder Baker, *Persons and Bodies: A Constitution View* (Cambridge: Cambridge University Press, 2000), chapter 8.
45. Lee M. Silver, *Remaking Eden* (New York: Avon, 1997).
46. See Svenaeus, *Phenomenological Bioethics*, chapter 6.

47. Baker, *Persons and Bodies*, chapter 4.
48. Svenaeus, *Phenomenological Bioethics*, chapter 2 and 3.
49. Jonas, *Technik, Medizin und Ethik*, chapter 10.
50. Heidegger, *Being and Time*, 45–51.
51. John P. Lizza, *Persons, Humanity, and the Definition of Death* (Baltimore: The Johns Hopkins University Press, 2006).
52. Ibid., 32–3, 163, 178–80.
53. Ibid., 63 ff.
54. See the early work of Hans Jonas, *The Phenomenon of Life* (New York: Harper & Row, 1966).
55. Lizza, *Persons, Humanity, and the Definition of Death*, 17 ff.
56. Svenaeus, *Phenomenological Bioethics*, chapter 7.
57. DeGrazia, *Human Identity and Bioethics*, chapter 4; and Younger, "The Definition of Death."
58. Heidegger, *Being and Time*, 238.
59. Sigmund Freud, "The Uncanny," in *Collected Papers*, vol. 4, trans. Joan Riviere (New York: Basic Books, 1959), 368–407.

Bibliography

Agamben, Giorgio. *The Time That Remains: A Commentary on the Letter to the Romans*, translated by Patricia Dailey. Stanford, CA: Stanford University Press, 2005.

Ahmed, Sara. *The Cultural Politics of Emotion*. New York: Routledge, 2004.

Aho, James, and Kevin Aho. *Body Matters: A Phenomenology of Sickness, Disease, and Illness*. Lanham: Lexington Books, 2008.

Aho, Kevin. *Heidegger's Neglect of the Body*. New York: State University of New York Press, 2007.

Aho, Kevin. "Heidegger, Ontological Death, and the Healing Professions." *Medicine, Health Care and Philosophy* 19, no. 1 (2016): 55–63.

Alleg, Henri. *The Question*. New York: George Braziller, 1958.

American Psychiatric Association. *Diagnostic and Statistical Manual of Mental Disorders*. American Psychiatric Association: Washington, DC, 2013.

Améry, Jean. "The Birth of Man from the Spirit of Violence: Frantz Fanon the Revolutionary." translated by Adrian Daub. *Wasafiri* 20, no. 44 (2005): 13–18.

Améry, Jean. *At the Mind's Limits*, translated by Sidney Rosenfeld and Stella P. Rosenfeld. Bloomington, IN: Indiana University Press, 1980.

Améry, Jean. *Jenseits von Schuld un Sühne*. Stuttgart: Klett-Cotta, 2002.

Amhed, Sara. *Queer Phenomenology: Orientations, Objects, Others*. Durham: Duke University Press, 2006.

Anscombe, Elisabeth. *Intention*. Cambridge: Harvard University Press, 2000.

Arendt, Hannah. *The Origins of Totalitarianism*. New York: Harcourt Books, 1966.

Aron, Lewis. "The Paradoxical Place of Enactment in Psychoanalysis: Introduction." *Psychoanalytic Dialogues* 13, no. 5 (2003): 623–31.

Arroll, M.A., and Senior, V. "Individuals' Experience of Chronic Fatigue Syndrome/ Myalgic Encephalomyelitis: An Interpretative Phenomenological Analysis." *Psychology and Health* 23, no. 4 (2008): 443–58.

Asad, Talal. *Formations of the Secular: Christianity, Islam, Modernity*. Stanford, CA: Stanford University Press, 2003.

Baker, Lynne Rudder. *Persons and Bodies: A Constitution View*. Cambridge: Cambridge University Press, 2000.

Barbaras, Renaud. "Affectivity and Movement: The Sense of Sensing in Erwin Straus." *Phenomenology and the Cognitive Sciences* 3 (2004): 215–28.

Barbaras, Renaud. "The Essence of Life: Drive or Desire?" In *Michel Henry. The Effects of Thought*, edited by Jeffrey Hanson, and Michael R. Kelly. London: Bloomsbury Press, 2012.

Bass, S. Jonathan. *Blessed Are the Peacemakers: Martin Luther King, Jr., Eight White Religious Leaders, and the "Letter from Birmingham Jail."* Baton Rouge: Louisiana State University Press, 2001.

Bauby, Jean-Dominique. *The Diving Bell and the Butterfly: A Memoir of Life in Death*, translated by Jeremy Leggatt. New York: Vintage Books, 1998.

Bauman, Zygmunt. *Collateral Damage: Social Inequalities in a Global Age.* Cambridge: Polity, 2011.

Bauman, Zygmunt. *Postmodern Ethics.* Oxford: Blackwell, 1993.

Beauvoir, Simone de. *The Second Sex*, translated by Constance Border and Sheila Malovany-Chevallier. New York: Knopf, 2010.

Beauvoir, Simone de, and Gisèle Halimi. *Djamila Boupacha*, translated by Peter Green. New York: Palgrave Macmillan, 1962.

Bellieni, Carlo Valerio. "Pain Assessment in Human Fetus and Infants." *The AAPS Journal* 14, no. 3 (2012): 456–61.

Benjamin, Jessica. "Where's the Gap and What's the Difference? The Relational View of Intersubjectivity, Multiple Selves, and Enactments." *Contemporary Psychoanalysis* 46, no. 1 (2010): 112–19.

Bergson, Henri. *Matter and Memory*, translated by Nancy Margaret Paul and W. Scott Palmer. New York: Zone Books, 1991.

Bernet, Rudolph. "Unconscious Consciousness in Husserl and Freud." In *The New Husserl: A Critical Reader*, edited by Donn Welton. Bloomington, IN: Indiana University Press, 2003.

Binswanger, Ludwig. "Über Psychotherapie." In *Ausgewählte Werke Band 3, Vorträge und Aufsätze*, edited by Max Herzog. Heidelberg: Verlag, 1994.

Bishop, Jeffrey P. *The Anticipatory Corpse: Medicine, Power, and the Care of the Dying.* Notre Dame: University of Notre Dame Press, 2011.

Black, David Allan. *Paul, Apostle of Weakness: Astheneia and its Cognates in the Pauline Literature.* New York: Peter Lang, 1984.

Bloechl, Jeffrey. "The Difficulty of Being: A Partial Reading of E. Levinas, De l'existence à l'existant." *European Journal of Psychotherapy, Counselling and Health* 7, nos. 1–2 (March–June 2005): 77–87.

Bornemark, Jonna. "Life Beyond Individuality: A-Subjective Experience in Pregnancy." In *Phenomenology of Pregnancy*, edited by Jonna Bornemark, and Nicholas Smith, 251–79. Huddinge: Södertön University Press, 2015.

Brooke, Roger. "Merleau-Ponty's Conception of the Unconscious." *African Journal of Psychology* 16 (1986): 126–30.

Butler, Judith. *Frames of War: When is Life Greivable?* New York: Verso Books, 2009.

Butler, Judith. *Precarious Life: The Powers of Mourning and Violence.* New York: Verso Books, 2001.

Buytendijk, Fredrick Jacobus Johannes. *Pain*, translated by Eda O'Shiel. London: Hutchinson, 1961.

Canguilhem, Georges. *The Normal and the Pathological.* New York: Zone Books, 2015.

Carel, Havi. *Illness: The Cry of the Flesh.* New York and London: Routledge, 2014.

Carel, Havi. *Phenomenology of Illness.* Oxford: Oxford University Press, 2015.

Carel, Havi. "The Philosophical Role of Illness." *Metaphilosophy* 45 (2014): 20–40.

Charon, Rita. *Narrative Medicine: Honoring the Stories of Illness*. Oxford: Oxford University Press, 2006.

Ciocan, Cristian. "Husserl's Phenomenology of Animality and the Paradoxes of Normality." *Human Studies* 40 (2017): 175–90.

Cioffi, Frank. *Wittgenstein on Freud and Frazier*. Cambridge: Cambridge University Press, 1998.

Cola, Brandon. "What Does Dabiq Do? ISIS Hermeneutics and Organizational Fractures within Dabiq Magazine." *Studies in Conflict & Terrorism* 40, no. 3 (June 2016): 173–90.

Cole, Jonathan. *Still Lives: Narratives of Spinal Cord Injury*. Cambridge: The MIT Press, 2004.

Conrad, Joseph. *The Secret Agent*. Oxford: Oxford University Press, 2008.

Conrad, Peter. *The Medicalization of Society: On the Transformation of Human Conditions into Treatable Disorders*. Baltimore: The Johns Hopkins University Press.

Cooper, Andrew. "Beyond Heidegger: From Ontology to Action." *Thesis Eleven* 40, no. 1 (2017): 90–105.

Cooper, Lesley. "Myalgic Encephalomyelitis and the Medical Encounter." *Sociology of Health and Illness* 19, no. 2 (1997): 186–207.

Couser, Thomas. *Recovering Bodies: Illness, Disability, and Life Writing*. Madison: University of Wisconsin Press, 1997.

Cox, Diane L. *Occupational Therapy and Chronic Fatigue Syndrome*. London: Whurr Publishers, 2000.

Dahlstrom, Daniel. "Heidegger's Method: Philosophical Concepts as Formal Indications." *The Review of Metaphysics* 47, no. 4 (1994): 775–95.

Dass, Ram. *Still Here: Embracing Aging, Changing, and Dying*. New York: Riverhead Books, 2001.

DeGrazia, David. *Human Identity and Bioethics*. Cambridge: Cambridge University Press, 2005.

DeLaure, Maralyn, and Bernard K. Duffy. "Martin Luther King, Jr. 1929–1968." In *American Voices: An Encyclopedia of Contemporary Orators*, edited by Bernard K. Duffy, and Richard W. Leeman, 258–69. Westport, CT: Greenwood Press, 2005.

Deleuze, Gilles. *Francis Bacon: The Logic of Sensation*, translated by Daniel W. Smith. Minneapolis: University of Minnesota Press, 2004.

DelVecchio Good, Mary-Jo, Paul E. Brodwin, Byron J. Good, and Arthur Kleinman, eds. *Pain as Human Experience: An Anthropological Perspective*. Berkeley: University of California Press, 1994.

Derrida, Jacques. *The Work of Mourning*, translated by Pascale-Anne Brault and Michael Naas. Chicago: The University of Chicago Press, 2001.

de Waal, Frans. *Primates and Philosophers: How Morality Evolved*. Princeton, NJ: Princeton University Press, 2006.

Dickson, Adele, Christina Knussen, and Paul Flowers. "'That Was My Old Life; It's Almost Like a Past-Life Now': Identity Crisis, Loss and Adjustment Amongst People Living with Chronic Fatigue Syndrome." *Psychology and Health* 23, no. 4 (2008): 459–76.

Dickson, Adele, Christina Knussen, and Paul Flowers. "Stigma and the Delegitimation Experience: An Interpretative Phenomenological Analysis of People

Living with Chronic Fatigue Syndrome." *Psychology and Health* 22, no. 7 (2007): 851–67.

Douglas, Guy. "Why Pains Are Not Mental Objects." *Philosophical Studies* 91 (1998): 127–48.

EEAS Strategic Planning. *A Secure Europe in a Better World: European Security Strategy Report* (December 12, 2003). European Union: Brussels.

El Damanhoury, Kareem, and Carol Winkler. "Picturing Law and Order: A Visual Framing Analysis of ISIS's *Dabiq* Magazine." *Arab Media and Society* (February 15, 2018). https://www.arabmediasociety.com/picturing-law-and-order-a-visual-framing-analysis-of-isiss-dabiq-magazine/

Eriksen, Thor Eirik, Anna Luise Kirkengen, and Arne Johan Vetlesen. "The Medically Unexplained Revisited." *Medicine Health Care and Philosophy* 16, no. 3 (2013): 587–600.

Eriksen, Thor Eirik, and Risør, Mette Bech. "What Is Called a Symptom?" *Medicine Health Care and Philosophy* 17, no. 1 (2014): 89–102.

Falke, Cassandra. *The Phenomenology of Love and Reading*. New York: Bloomsbury Press, 2017.

Fanon, Frantz. *Black Skin, White Masks*, translated by Richard Philcox. New York: Grove Press, 2008.

Fanon, Frantz. *Toward the African Revolution*, translated by Haakon Chevalier. New York: Grove Press, 1967.

Fanon, Frantz. *The Wretched of the Earth*, translated by Richard Philcox. New York: Grove Press, 2004.

Farquhar, Judith. *Knowing Practice: The Clinical Encounter of Chinese Medicine*. New York: Routledge, 1994.

Flensner, Gullvi, Anna-Christina Ek, and Olle Soderhamn. "Lived Experience of MS-Related Fatigue—A Phenomenological Interview Study." *International Journal of Nursing Studies* 40 (2003): 707–17.

Foucault, Michel. "Introduction." In *The Normal and the Pathological*. New York: Zone Books, 2015.

Francis, Claude, and Ferdinande Gontier. *Simone de Beauvoir: A Life, A Love Story*. London: Vermilion Books, 1988.

Frank, Arthur W. *At the Will of the Body*. Boston: Houghlin Mifflin, 1991.

Frank, Arthur W. *The Wounded Storyteller: Body, Illness, and Ethics*. Chicago: The University of Chicago Press, 2013.

Frank, Didier. *Flesh and Body: On the Phenomenology of Husserl*, translated by J. Rivera and S. Davidson. London: Bloomsbury Press, 2014.

Freud, Sigmund. "The Ego and the Id." In *The Essentials of Psycho-analysis*, edited by Anna Freud. London: Vintage Books, 2005.

Freud, Sigmund. *Inhibitions, Symptoms and Anxiety*, translated by Alix Strachey. London: Hogarth Press, 1936.

Freud, Sigmund. *Introductory Lectures on Psycho-Analysis*, translated by James Strachey. New York: W. W. Norton and Company, 1989.

Freud, Sigmund. "Recommendations to Physicians Practicing Psycho-Analysis." In *The Standard Edition of the Complete Psychological Works of Sigmund Freud XII*, edited by James Strachey. London: Vintage Books, 1999.

Freud, Sigmund. "The Uncanny." In *Collected Papers*, Vol. 4, translated by Joan Riviere, 368–407. New York: Basic Books, 1959.

Freund, Julia, et al. "The Emergence of Individuality in Genetically Identical Mice." *Science* 340, no. 6133 (May 2013): 756–9.

Fricker, Miranda. *Epistemic Injustice: Power and the Ethics of Knowing*. Oxford: Oxford University Press, 2007.

Fuchs, Thomas. "Body Memory and the Unconsciousness." In *Founding Psychoanalyses phenomenologically*, edited by Dieter Lohmar, and Jagna Brudzínska. Dordrecht: Verlag, 2012.

Fuchs, Thomas. "Corporealized and Disembodied Minds—A Phenomenological View of the Body in Melancholia and Schizophrenia." *Philosophy, Psychiatry and Psychology* 12 (2005): 95–107.

Fuchs, Thomas. "The Phenomenology of Affectivity." In *The Oxford Handbook of Philosophy and Psychiatry*, edited by K.W.M. Fulford, Martin Davies, Richard G.T. Gipps, George Graham, John Z. Sadler, Giovanni Stanghellini, and Tim Thornton, 612–31. Oxford: Oxford University Press, 2013.

Fuchs, Thomas. "Psychotherapy of the Lived Space: A Phenomenological and Ecological Concept." *American Journal of Psychotherapy* 61, no. 4 (2007).

Gallagher, Shaun. "Body Image and Body Schema: A Conceptual Clarification." *The Journal of Mind and Behavior* 7, no. 4 (Autumn 1986): 541–54.

Gallagher, Shaun. *Enactivist Interventions*. Oxford: Oxford University Press, 2017.

Gallagher, Shaun. *How the Body Shapes the Mind*. Oxford: Oxford University Press, 2005.

Gallagher, Shaun. *Phenomenology*. Basingstoke: Palgrave MacMillan, 2012.

Garro, Linda C. "Chronic Illness and the Construction of Narratives." In *Pain as Human Experience: An Anthropological Perspective*, edited by DelVecchio Good et al, 100–37, 1994. Berkeley: University of California Press.

Gawande, Atul. *Being Mortal: Illness, Medicine and What Matters in the End*. London: Profile Books, 2014.

Geniusias, Saulius. "Pain and Intentionality." In *Perception, Affectivity, and Volition in Husserl's Phenomenology*, edited by Roberto Walton, Shigeru Taguchi, and Roberto Rubio, 113–33. Cham: Springer Verlag, 2017.

Geniusias, Saulius. "Phenomenology of Chronic Pain: De-personalization and Re-personalization." In *Meanings of Pain*, edited by Simon van Rysewyk, 147–64. Cham: Springer Verlag, 2016.

Gilroy, Paul. "Fanon and Améry: Theory, Torture and the Prospect of Humanism." *Theory, Culture & Society* 27, nos. 7–8 (2010): 16–32.

Goldstein, Kurt. *The Organism: A Holistic Approach to Biology Derived from Pathological Data in Man*. New York: Zone Books, 1995.

Good, Byron J. *Medicine, Rationality, and Experience: An Anthropological Perspective*. Cambridge: Cambridge University Press, 1994.

Graça Marcel. "Promotion and Protection of the Rights of Children Impact of Armed Conflict on Children." (August 26, 1996). Report to the UN General Assembly. 9.

Grahek, Nikola. *Feeling Pain and Being in Pain*. Cambridge: The MIT Press, 2001.

Grosz, Elizabeth. *Volatile Bodies: Toward a Corporeal Feminism*. Bloomington, IN: Indiana University Press, 1994.

Grüny, Christian. "Vom Nutzen und Nachteil des Schmerzes für das Leben." In *Schmerz als Grenzerfahrung*, edited by Rainer M.E. Jacobi, and Bernhard Marx, 39–57. Leipzig: Evangelische Verlagsanstalt, 2011.

Grüny, Christian. *Zerstörte Erfahrung: Eine Phänomenologie des Schmerzes.* Würzburg: Königshausen & Neumann, 2004.

Grüny, Christian. "Zur Logik der Folter." In *Gewalt-Verstehen*, edited by Burkhard Liebsch, and Dagmar Mensink, 79–115. Berlin: Verlag, 2003.

Gschwandtner, Christina M. *Degrees of Givenness: On Saturation in Jean-Luc Mairon.* Bloomington, IN: Indiana University Press, 2014.

Hegghammer, Thomas. *Un-Inspired.* Jihadica: Documenting the Global Jihad (July 6, 2010).

Heidegger, Martin. *Basic Concepts of Aristotelian Philosophy*, translated by R.D. Metcalf and M.B. Tanzer. Bloomington, IN: Indiana University Press, 1993.

Heidegger, Martin. *Being and Time*, translated by John Macquarrie and Edward Robinson. New York: Harper Perennial, 1962.

Heidegger, Martin. *Being and Time*, translated by Joan Stambaugh. Albany: State University of New York Press, 2010.

Heidegger, Martin. *What is Called Thinking?* translated by J. Glenn Gray. New York: Perennial Library, 2004.

Heidegger, Martin. *Zollikoner Seminare*, edited by Medard Boss. Frankfurt am Main: Vittorio Klostermann, 1987.

Henry, Michel. *Incarnation: A Philosophy of the Flesh*, translated by Karl Hefty. Evanston, IL: Northwestern University Press, 2015.

Henry, Michel. "Phenomenology of Life." *Angelaki Journal of the Theoretical Humanities* 8, no. 2 (June 2010): 97–110.

Hofmann, Bjørn, and Fredrik Svenaeus. "How Medical Technologies Shape the Experience of Illness." *Life Sciences, Society and Policy* 14, no. 3 (2018): 1–11.

Hughes, Julian, C., Stehpen J. Louw, and Steven R. Sabat, eds. *Dementia: Mind, Meaning and the Person.* Oxford: Oxford University Press, 2006.

Husserl, Edmund. *Analyses Concerning Passive and Active Synthesis*, translated by Anthony J. Steinbock. Dordrecht: Kluwer Academic Publishers, 2001.

Husserl, Edmund. *The Crisis of European Sciences and Transcendental Phenomenology*, translated by David Carr. Evanston, IL: North Western University Press, 1970.

Husserl, Edmund. *Experience and Judgment: Investigations in a Genealogy of Logic*, translated by James Churchill and Karl Ameriks. Evanston, IL: Northwestern University Press, 1973.

Husserl, Edmund. *Ideas Pertaining to a Pure Phenomenology and to a Phenomenological Philosophy*, Vol. I, translated by Fred Kersten. The Hague: Martinus Niejhoff, 1983.

Husserl, Edmund. *Ideas Pertaining to a Pure Phenomenology and to a Phenomenological Philosophy*, *Second Book: Studies in the Phenomenology of Constitution*, translated by Richard Rojcewicz and André Schuwer. Dordrecht: Kluwer Academic Publishers, 1989.

Irigaray, Luce. *An Ethics of Sexual Difference*, translated by Carolyn Burke and Gillian C. Gill. Ithaca: Cornell University Press, 1993.

Jackson, J.E. " 'After a While No One Believes You': Real and Unreal Pain." In *Pain as Human Experience: An Anthropological Perspective*, edited by DelVecchio Good et al., 138–68. Berkley: University of California Press, 1994.

Jaspers, Karl. *General Psychopathology*, Vol. 2. Baltimore: Johns Hopkins Press, 1997.

Johnson, Anne. "Exploring the Experiences and Occupations of Men with Chronic Fatigue Syndrome/Myalgic Encephalomyelitis (CFS/ME) Using a

Gadamerian Interpretive Phenomenological Framework." PhD diss., University of the West of England, 2017, http://eprints.uwe.ac.uk/29916

Jonas, Hans. "Causality and Perception." *The Journal of Philosophy* 47, no. 11 (May 1950): 319–24.

Jonas, Hans. *The Imperative of Responsibility: In Search for an Ethics for the Technological Age.* Chicago: Chicago University Press, 1984.

Jonas, Hans. *The Phenomenon of Life.* New York: Harper & Row, 1966.

Jonas, Hans. *Technik, Medizin und Ethik.* Frankfurt am Main: Suhrkamp Verlag, 1987.

Kalanithi, Paul. *When Breath Becomes Air.* New York: Random House, 2016.

Kearney, Richard, and Brian Treanor, eds. *Carnal Hermeneutics.* New York: Fordham University Press, 2015.

King, Martin Luther. *Where Do We Go from Here: Chaos or Community?* Boston: Beacon Press, 1968.

King, Martin Luther. *Why We Can't Wait.* New York: Harper & Row, 1964.

Klein, Melanie. *Envy and Gratitude and Other Works 1946–1963.* London: Vintage Books, 1997.

Kleinman, Arthur. *The Illness Narratives: Suffering, Healing and the Human Condition.* New York: Basic Books, 1988.

Kleinman, Arthur. "The Social Course of Chronic Illness." In *Chronic Illness: From Experience to Policy*, edited by S.K. Toombs, David Barnard, and Ronald A. Carson, 176–88. Bloomington, IN: Indiana University Press, 1995.

Krüger's, Hans-Peter. "Persons and Their Bodies: The *Körper/Leib* Distinction and Helmuth Plessner's Theories of Ex-centric Positionality and *Homo asconditus.*" *Journal of Speculative Philosophy* 24, no. 3 (2010): 256–74.

Lacan, Jacques. *On Feminine Sexuality: The Limits of Love and Knowledge, Book XX: Encore, 1972–1973*, translated by Bruce Fink. New York: W. W. Norton and Company, 1999.

Lanzoni, Susan. "Existential Encounter in the Asylum: Ludwig Binswanger's 1935 case of Hysteria." *History of Psychiatry* 15, no. 3: 285–304.

Larsson, Katharine M. "Understanding the Lived Experience of Patients who Suffer from Medically Unexplained Physical Symptoms Using a Rogerian Perspective." PhD diss., Boston College, William F. Connell School of Nursing, USA, 2008.

Leder, Drew. *The Absent Body.* Chicago and London: The University of Chicago Press, 1990.

Leder, Drew. *The Distressed Body: Rethinking Illness, Imprisonment and Healing.* Chicago: The University of Chicago Press, 2016.

Leder, Drew. *Spiritual Passages: Embracing Life's Sacred Journey.* New York: Penguin Books and Tarcher, 1997.

Legrand, Dorothée, and Dylan Trigg, ed. *Unconsciousness between Phenomenology and Psychoanalysis* (Contributions to Phenomenology 88). Cham: Springer Verlag, 2017.

LeShan, Lawrence. "The World of the Patient of Severe Pain of Long Duration." *Journal of Chronic Diseases* 17 (1964): 119–26.

Levinas, Emmanuel. *Alterity and Transcendence*, translated by Michael B. Smith. London: The Athelone Press, 1999.

Levinas, Emmanuel. *Existence and Existents*, translated by Alphonso Lingis. Dordrecht, Boston and London: Kluwer Academic Publishers, 1978.

Levinas, Emmanuel. *Otherwise than Being, or Beyond Essence*, translated by Alphonso Lingis. Pittsburgh, PA: Duquesne University Press, 1998.

Levinas, Emmanuel. "There Is: Existence without Existents." In *The Levinas Reader*, edited by Sean Hand. Oxford: Blackwell, 1989.

Levinas, Emmanuel. *Totality and Infinity*, translated by Alphonso Lingis. Dordrecht: Kluwer Academic Publishers, 1991.

Levine, Joseph. "Materialism and Qualia: The Explanatory Gap." *Pacific Philosophical Quarterly* 64 (1983): 354–61.

Levine, Lauren. "Into Thin Air: The Co-Construction of Shame, Recognition and Creativity in an Analytic Process." *Psychoanalytic Dialogues* 22 (2012): 456–71.

Lewis, Michael, and Tanja Staehler. *Phenomenology: An Introduction*. London: Continuum, 2010.

Lintott, Sheila. "The Sublimity of Gestating and Giving Birth: Toward a Feminist Conception of the Sublime." In *Philosophical Inquires into Pregnancy, Childbirth, and Mothering*, edited by Sheila Lintott, and Maureen Sander-Staudt, 237–50. New York: Routledge, 2012.

Lizza, John P. *Persons, Humanity, and the Definition of Death*. Baltimore: The Johns Hopkins University Press, 2006.

Lonardi, Cristina. "The Passing Dilemma in Socially Invisible Diseases: Narratives on Chronic Headache." *Social Science and Medicine* 65 (2007): 1619–29.

Lurija, A.R. *The Man with a Shattered World: A History of a Brain Wound*. Harmondsworth: Penguin Books, 1975.

MacIntyre, Alasdair. *Dependent Rational Animals: Why Human Beings Need the Virtues*. Chicago: Open Court, 2001.

Marcel, Gabriel. *Being and Having*, translated by Katharina Farrer. Westminster: Dacre Press, 1949.

Marcel, Gabriel. *The Mystery of Being: Volume I: Reflection and Mystery*, translated by Rene Hague. Indiana: St. Augustine's Press, 2001.

Marcel, Gabriel. *The Philosophy of Existentialism*, translated by Manya Harari. New York: Citadel Press, 1984.

Marion, Jean-Luc. *Being Given: Toward a Phenomenology of Givenness*, translated by Jeffrey L. Kosky. Stanford, CA: Stanford University Press, 2002.

Marion, Jean-Luc. *The Erotic Phenomenon*, translated by Stephen Lewis. Chicago: The University of Chicago Press, 2007.

Marion, Jean-Luc. *In Excess: Studies in Saturated Phenomena*, translated by Robyn Horner and Vincent Berraud. New York: Fordham University Press, 2002.

Marion, Jean-Luc. *Negative Certainties*, translated by Stephen E. Lewis. Chicago: The University of Chicago Press, 2015.

McMahan, Jeff. "Death, Brain Death, and Persistent Vegetative State." In *A Companion to Bioethics*, edited by Helga Kushe, and Peter Singer, 286–98. London: Wiley-Blackwell, 2009.

McWilliams, Nancy. *Psychoanalytic Diagnosis*. New York: The Guilford Press, 2011.

Melzack, Ronald, and Patrick D. Wall. *The Challenge of Pain*. London: Penguin Books, 1988.

Merleau-Ponty, Maurice. *Child Psychology and Pedagogy: The Sorbonne Lectures 1949–1952*, translated by Talia Welsh. Evanston, IL: Northwestern University Press, 2010.

Merleau-Ponty, Maurice. *Institution and Passivity—Course Notes from the Collège de France (1954–1955)*. Evanston, IL: Northwestern University Press, 2016.

Merleau-Ponty, Maurice. *Phenomenology of Perception*, translated by Colin Smith. London: Routledge, 1992.

Merleau-Ponty, Maurice. *Phenomenology of Perception*, translated by Donald A. Landes. London: Routledge, 2012.

Merleau-Ponty, Maurice. *The Primacy of Perception and other Essays*, translated and edited by J.M. Edie. Evanston, IL: Northwestern University Press, 1964.

Merleau-Ponty, Maurice. *Signs*, translated R.C. McCleary. Evanston, IL: Northwestern University Press, 1964.

Merleau-Ponty, Maurice. *The Structure of Behaviour*, translated by Alden L. Fisher. Pittsburgh, PA: Duquesne University Press, 2008.

Merleau-Ponty, Maurice. *Themes from the Lectures at Collége de France 1952–1960*, translated by John O'Neill. Evanston, IL: Northwestern University Press, 1979.

Merleau-Ponty, Maurice. *The Visible and the Invisible*, translated by Alphonso Lingis. Evanston, IL: Northwestern University Press, 1968.

Milella, Pietro Paolo. "Nature and Phenomenology of Fatigue." In *His Fatigue and Corrosion in Metals*. Milano: Springer, 2013.

Moran, Dermot. "Husserl's Layered Concept of the Human Person: Conscious and Unconscious." In *Unconsciousness between Phenomenology and Psychoanalysis (Contributions to Phenomenology 88)*, edited by Dorothée Legrand, and Dylan Trigg, 3–25. Cham: Springer Verlag, 2017.

Morris, Katherine J. "Chronic Pain in Phenomenological/Anthropological Perspective." In *The Phenomenology of Embodied Subjectivity*, edited by Rasmus Thybo Jansen, and Dermot Moran, 167–84. Cham and Heidelberg: Springer Verlag, 2013.

Morris, Katherine J. "The Phenomenology of Clumsiness." In *Sartre on the Body*, edited by Morris, 161–82, Basingstoke and New York: Palgrave Macmillan, 2010.

Morris, Katherine J., ed. *Sartre on the Body*. Basingstoke and New York: Palgrave Macmillan, 2010.

Morris, Katherine J. *Starting with Merleau-Ponty*. London and New York: Continuum, 2012.

Moss, Candida R. "Christly Possession and Weakened Bodies: Reconsideration of the Function of Paul's Thorn in the Flesh." *Journal of Religion, Disability & Health* 16 (2012): 319–33.

Myhill, Sarah. *Diagnosis and Treatment of Chronic Fatigue Syndrome: Mitochondria, Not Hypochondria*. London: Hammersmith, 2014.

Nagel, Thomas. *Mind & Cosmos: Why the Materialist Neo-Darwinian Conception of Nature is Almost Certainly False*. Oxford: Oxford University Press, 2012.

Nagel, Thomas. "What Is It Like to Be a Bat?" *The Philosophical Review* 83, no. 4 (1974): 435–50.

Nietzsche, Friedrich. *The Gay Science: With a Prelude in German Rhymes and an Appendix of Songs*, translated by Josefine Nauckhoff. Cambridge: Cambridge University Press, 2001.

Noë, Alva. *Action in Perception*. London: The MIT Press, 2004.

O'Connell, Mark. *To Be a Machine: Adventures Among Cyborgs, Utopians, Hackers, and the Futurists Solving the Modest Problem of Death*. New York: Doubleday, 2017.

Oksala, Johanna. "What Is Feminist Phenomenology? Thinking Birth Philosophically." *Radical Philosophy* 26 (July/August 2004): 16–22.

Olivier, Abraham. *Being in Pain*. Frankfurt am Main: Peter Lang, 2007.

Olivier, Abraham. "When Pains are Mental Objects." *Philosophical Studies* 115 (2003): 33–53.

OMEGA. *Poetry from the Bed*. Oxfordshire: Oxfordshire Myalgic Encephalomyelitis Group for Action, 2012.

Orange, Donna M. *The Suffering Stranger: Hermeneutics for Everyday Clinical Practice*. New York and London: Routledge, 2011.

Parfit, Derek. *Reasons and Persons*. Oxford: Oxford University Press, 1984.

Parnas, Josef, and Dan Zahavi. "The Link: Philosophy—Psychopathology—Phenomenology." In *Exploring the Self. Philosophical and Psychopathological Perspectives on Self-Experience*, edited by Dan Zahavi, 1–16. Philadelphia: John Benjamins Publishing, 2000.

Pascal, Blaise. *Pensées*. Project Gutenberg. www.gutenberg.org/files/18269/18269-h/18269-h.htm

Peckitt, Michael Gillan. "Resisting Sartrean Pain." In *Sartre on the Body*, edited by Morris, 120–9, 2010. Basingstoke and New York: Palgrave Macmillan.

Perkins, Judith. *The Suffering Self: Pain and Narrative Representation in the Early Christian Era*. London and New York: Routledge, 1995.

Phillips, James. "Merleau-Ponty's Non-Verbal Unconsciousness." In *Unconsciousness between Phenomenology and Psychoanalysis (Contributions to Phenomenology 88)*, edited by Dorothée Legrand, and Dylan Trigg, 75–95. Cham: Springer Verlag, 2017.

Pinker, Steven. *The Better Angels of Our Nature: Why Violence Has Declined*. New York: Viking, 2011.

Plaut, W. Gunther, ed. *The Torah: A Modern Commentary*. New York: Union of American Hebrew Congregations, 1981.

Plessner, Helmuth. *Die Stufen des Organischen und der Mensch*. Berlin: de Gruyter, 1975.

Plügge, Herbert. *Der Mensch und sein Leib*. Tübingen: Niemeyer, 1967.

Ponsi, Maria. "The Evolution of Psychoanalytic Thought: Acting Out and Enactment." *The Italian Psychoanalytic Annual* (2013): 16.

Pradeu, Thomas. *The Limits of the Self: Immunology and Biological Identity*, translated by Elizabeth Vitanza. Oxford: Oxford University Press, 2012.

Rawnsley Andrew C. "Practice and Givenness: The Problem of 'Reduction' in the Work of Jean-Luc Marion." *New Blackfriars* 88, no. 1018. Oxford: Blackwell, 2007.

Reindal, Solveig Magnus. "Disability, Gene Therapy and Eugenics—a Challenge to John Harris." *Journal of Medical Ethics* 26, no. 2 (2000): 89–94.

Rejali, Darius. *Torture and Democracy*. Princeton, NJ: Princeton University Press, 2007.

Ricoeur, Paul. *The Conflict of Interpretations*, edited by John Ihde. New York and London: Continuum, 1989.

Ricoeur, Paul. *Oneself as Another*, translated by Kathleen Blamey. Chicago: The University of Chicago Press, 1992.

Riemer, Jack. "Perlman makes his music the hard way." *Houston Chronicle* (February 10, 2001), www.chron.com/life/houston-belief/article/Perlman-makes-his-music-the-hard-way-2009719.php

Robinson, John A.T. *The Body: A Study in Pauline Theology*. London: SCM Press, 1952.

Rochat, Philippe. *Others in Mind: Social Origins of Self-Consciousness*. Cambridge: Cambridge University Press, 2009.

Romanyshyn, D. "Phenomenology and Psychoanalysis." *Psychoanalytic Review* 64 (1977): 211–23.

Romdenh-Romluc, Komarine. "Merleau-Ponty and the Power to Reckon with the Possible." In *Reading Merleau-Ponty*, edited by Thomas Baldwin, 44–58. Abingdon and Oxon: Routledge, 2007.

Ruin, Hans. "Sacrificial Subjectivity: Faith and Interiorization of Cultic Practice in the Pauline Letters." In *Philosophy and the End of Sacrifice*, edited by P. Jackson, and A.P. Sjödin. Sheffield: Equinox, 2016.

Russell, Ronald. "Redemptive Suffering and Paul's Thorn in the Flesh." *Journal of Evangelical Theological Society* 39 (1996): 559–70.

Russon, John. *Human Experience*. Albany: SUNY Press, 2003.

Sacks, Oliver. *The Man Who Mistook His Wife for a Hat and Other Clinical Tales*. New York: Harper Perennial, 1987.

Sartre, Jean-Paul. *Being and Nothingness*, translated by Hazel E. Barnes. New York: Washington Square Press, 1993.

Sartre, Jean-Paul. *Being and Nothingness an Essay on Phenomenological Ontology*, translated by Hazel E. Barnes. London: Routledge, 1986.

Sartre, Jean-Paul. *Critique of Dialectical Reason*, translated by Alan Sheridan-Smith. New York: Verso Books, 2004.

Sass, Louis A. *Madness and Modernism*. New York: Basic Books, 1992.

Sass, Louis A. "Self and World in Schizophrenia: Three Classic Approaches." *Philosophy, Psychiatry, & Psychology* 8, no. 4 (2001): 251–70.

Scarry, Elaine. *The Body in Pain: The Making and the Unmaking of the World*. Oxford: Oxford University Press, 1985.

Serrano de Haro, Agustín. "Pain Experience and Structures of Attention: A Phenomenological Approach." In *Meanings of Pain*, edited by Simon van Rysewyk, 165–80. Cham: Springer Verlag, 2016.

Sherman, Nancy. *The Untold War: Inside the Hearts, Minds, and Souls of Our Soldiers*. New York: W. W. Norton and Company, 2010.

Sigurdson, Ola. "The Body of Illness: Narrativity, Embodiment and Relationality in Doctoring and Nursing." English translation of "Sjukdomens kropp: Narrativitet, kroppslighet och relationalitet i medicinsk praktik och omvårdnad." *Kritisk forum for praktisk teologi* 31, no. 123 (2011): 6–22.

Sigurdson, Ola. "Existential Health: Philosophical and Historical Perspectives." *LIR Journal* 6 (2016): 7–23.

Sigurdson, Ola. *Heavenly Bodies: Incarnation, the Gaze, and Embodiment in Christian Theology*, translated by Carl Olsen. Grand Rapids: Eerdmans, 2016.

Silver, Lee M. *Remaking Eden*. New York: Avon, 1997.

Slim, Hugo. *Killing Civilians: Method, Madness and Morality in War*. Oxford: Oxford University Press, 2010.

Smith, Nicholas. "Phenomenology of Pregnancy: A Cure for Philosophy?" In *Phenomenology of Pregnancy*, edited by Jonna Bornemark, and Nicholas Smith, 15–49. Huddinge: Södertön University Press, 2015.

Sobchack, Vivian. " 'Choreography for One, Two, and Three Legs' (A Phenomenological Meditation in Movements)." *Topoi* 24 (2005): 55–66.

Staudigl, Michael. *Phenomenologies of Violence*. Leiden: Brill, 2014.

Staudigl, Michael. "Racism: On the Phenomenology of Embodied Desocialization." *Continental Philosophy Review* 45 (2012): 23–39. 28.

Steinbock, Anthony J. *Home and Beyond: Generative Phenomenology after Husserl*. Evanston, IL: Northwestern University Press, 1995.

Steinbock, Anthony J. *Moral Emotions: Reclaiming the Evidence of the Heart*. Evanston, IL: Northwestern University Press, 2014.

Stern, Daniel N. *The Interpersonal World of the Infant*. London: Karnac Books, 1985.

Stoller, Silvia. "Gender and Anoymous Temporality." In *Time in Feminist Phenomenology*, edited by Christina Schües, Dorothea Olkowski, and Helen Fielding, 79–90. Bloomington, IN: Indiana University Press, 2011.

Straus, Erwin. *Vom Sinn der Sinne: Ein Beitrag zur Grundlegung der Psychologie* (2nd ed.). Berlin: Springer, 1956.

Straus, Joseph N. *Extraordinary Measures: Disability in Music*. Oxford: Oxford University Press, 2011.

Stryk, Lucien, ed. *World of the Buddha*. New York: Grove Press, 1968.

Stumpf, Carl. *Gefühl und Gefühlsempfindung*. Leipzig: Barth, 1928.

Summa, Michela. "The Disoriented Self. Layers and Dynamics of Self-Experience in Dementia and Schizophrenia." *Phenomenology and the Cognitive Sciences* 13, no. 3 (2014): 477–96.

Svenaeus, Fredrik. *The Hermeneutics of Medicine and the Phenomenology of Health: Steps Towards a Philosophy of Medical Practice*. Dordrecht: Kluwer Academic Publishers, 2000.

Svenaeus, Fredrik. *Phenomenological Bioethics: Medical Technologies, Human Suffering, and the Meaning of Being Alive*. London: Routledge, 2017.

Taylor, Charles. *A Secular Age*. Cambridge, MA: The Belknap Press of Harvard University Press, 2007.

Tellenbach, Hubertus. "The Phenomenology of States of Health and Its Consequences for the Physician." *Morality Within the Life—and Social World* 22 (1987): 371–81.

Toombs, S. Kay. *The Meaning of Illness: A Phenomenological Account of the Different Perspectives of Physician and Patient*. Dordrecht: Kluwer Academic Publishers, 1992.

Waldenfels, Bernhard. *Phenomenology of the Alien: Basic Concepts*, translated by Alexander Kozin and Tanja Stähler. Evanston, IL: Northwestern University Press, 2011.

Ware, Norma C. "Suffering and the Social Construction of Illness: The Delegitimation of Illness Experience in Chronic Fatigue Syndrome." *Medical Anthropology Quarterly New Series* 6, no. 4 (December 1992): 347–61.

Weber, Elisabeth. *Kill Boxes: Facing the Legacy of US-Sponsored Torture, Indefinite Detention, and Drone Warfare*. New York: Punctum Books, 2017.

Welsh, Talia. "The Order of Life: How Phenomenologies of Pregnancy Revise and Reject Theories of the Subject." In *Coming to Life: Philosophies of Pregnancy, Childbirth, and Mothering*, edited by Sarah LaChance Adame, and Caroline Lundquist, 283–99. New York: Fordham University Press, 2013.

Welton, Donn, ed. *The New Husserl: A Critical Reader*. Bloomington, IN: Indiana University Press, 2003.

Wendell, Susan. *The Rejected Body: Feminist Reflections on Disability*. New York: Routledge, 1996.

Winkler, Carol, Kareem El Damanhoury, Aaron Dicker, and Anthony F. Lemieux. "The Medium Is Terrorism: Transformation of the About to Die Trope in Dabiq." *Terrorism and Political Violence* (2017): 1–20.

Winnicott, Donald W. *Playing and Reality*. London: Routledge, 1971.

Winnicott, Donald W. "The Theory of the Parent-Infant Relationship." *The International Journal of Psychoanalysis* 41 (1960): 585–95.

Withy, Katherine. *Heidegger on Being Uncanny*. Cambridge: Harvard University Press, 2015.

World Health Organization (WHO). *Dementia: Fact Sheet*, 2016. www.who.int/mediacentre/factsheets/fs362/en/

Young, Iris Marion. "Pregnant Embodiment: Subjectivity and Alienation." In *On Female Body Experience: 'Throwing Like a Girl' and Other Essays*, 46–61. Oxford: Oxford University Press, 2005.

Young, Iris Marion. "Throwing Like a Girl: A Phenomenology of Feminine Body Comportment Motility and Spatiality." *Human Studies* 3 (1980): 137–56.

Younger, Stuart J. "The Definition of Death." In *The Oxford Handbook of Bioethics*, edited by Bonny Steinbock, 285–303. Oxford: Oxford University Press, 2007.

Zahavi, Dan. *Subjectivity and Selfhood: Investigating the First-Person Perspective*. Cambridge: The MIT Press, 2005.

Zaner, Richard M. *The Context of Self*. Athens: Ohio University Press, 1981.

Contributors

Kevin Aho is Professor and Chair of the Department of Communication and Philosophy at Florida Gulf Coast University. He has published widely in the areas of existentialism, phenomenology, hermeneutics and the philosophy of medicine. He is the author of *Body Matters: A Phenomenology of Sickness, Illness, and Disease* (with James Aho, 2008), *Heidegger's Neglect of the Body* (2009), *Existentialism: An Introduction* (2014) and editor of *Existential Medicine: Essays on Health and Illness* (2018).

Espen Dahl is Professor of Systematic Theology at UiT-The Arctic University of Norway. His research interests mainly focus on the intersection between twentieth-century philosophy (phenomenology and ordinary language philosophy) and theology. His publications include *Stanley Cavell, Religion, and Continental Philosophy* (Indiana University Press 2014); *In Between. The Holy Beyond Modern Dichotomies* (Vandenhoek & Ruprecht 2011); *The Holy and Phenomenology. Religious Experience after Husserl* (SCM Press 2010). Dahl has published numerous articles on theology and philosophy, such as "Job and the Problem of Physical Pain—a Phenomenological Reading," *Modern Theology* 2016, 32 (1); and "Humility and Generosity: On the Horizontality of the Divine Givenness," *Neue Zeitschrift für systematische Theologie und Religionsphilosophie*, 55 (nr 3) (2013).

Thor Eirik Eriksen has a PhD in Philosophy and holds a position as senior adviser at The University Hospital of North Norway. His main research interests are philosophy of science, existential philosophy, phenomenology and the borderland between philosophy and medicine. He has been a contributing author on such articles as: "At the Borders of Medical Reasoning: Aetiological and Ontological Challenges of Medically Unexplained Symptoms" in *Philosophy, Ethics and Humanities in Medicine* (in press), "The Medically Unexplained Revisited" in *Medicine Healthcare and Philosophy* (2012), "Patients' 'Thingification,' Unexplained Symptoms and Response-ability in the Clinical Context" (2016).

Cassandra Falke is a Professor of English Literature at UiT-The Arctic University of Norway. Her books include *Intersections in Christianity and Critical Theory* (ed. Palgrave 2010), *Literature by the Working Class: English Autobiography, 1820–1848* (Cambria, 2013) and, most recently *The Phenomenology of Love and Reading* (Bloomsbury, 2016). She has also authored articles about Wordsworth, Byron, Coleridge, liberal arts education, contemporary phenomenology and the portrayal of violence in literature.

Ståle Finke is Professor of Philosophy at NTNU in Norway. He has written about aesthetics and political philosophy in the tradition from Kant and German Idealism, as well as phenomenology, critical theory and hermeneutics. Among Finke's publications are "The Imagination of Cultural Modernity: Heidegger and Cassirer in Davos," "Concepts and Intuitions: Adorno after the Linguistic Turn" and "The Hermeneutics of the Image: Gadamer and the Language of Art."

Christian Grüny teaches philosophy at the University Witten/Herdecke. He has taught and pursued his research at the universities of Witten/ Herdecke and Darmstadt and also at the Art Academy Düsseldorf and the Max Planck Institute for Empirical Aesthetics in Frankfurt. His areas of research are aesthetics, the philosophy of music, phenomenology and the philosophy of culture.

Drew Leder is a Professor of Philosophy at Loyola University. He is the author of many books, both scholarly and popular, including *The Distressed Body: Rethinking Illness, Imprisonment, and Healing* (University of Chicago 2016), *The Body in Medical Thought and Practice* (editor, Kluwer Academic Press, 1992) and *The Absent Body* (University of Chicago Press, 1990).

James McGuirk is Professor of Philosophy and former Head of the Centre for Practical Knowledge at Nord University, Norway. He has authored many journal articles and book chapters on themes and figures in the phenomenological tradition. His most recent publication is *Eros, Otherness, Tyranny: The Indictment and Defence of the Philosophical Life in Plato, Nietzsche, and Lévinas* (2017). His current research interests include the Philosophy of practical knowledge, Improvisation and practice and critical thinking and values in schools.

Alexandra Megearu is a PhD candidate in the Comparative Literature Program at the University of California, Santa Barbara. Her research focuses on the intersection between postcolonial and diasporic literature, feminist theory, affect studies, and continental philosophy. She is also a visual artist and writer.

Katherine J. Morris is a fellow in philosophy at Mansfield College, Oxford University, UK. Her books include *Descartes' Dualism* (with

Gordon Baker, Routledge, 1996), *Sartre* (Blackwell, 2008) and *Starting with Merleau-Ponty* (Continuum, 2012). She has published widely on Descartes, Wittgenstein, Sartre and Merleau-Ponty and has further research interests in medical anthropology.

Ola Sigurdson is a Professor of Systematic Theology at the Department of Literature, History of Ideas, and Religion, University of Gothenburg, Sweden. He is the author of more than 20 books in Swedish and English ranging from theology and contemporary continental philosophy to medical humanities. His most recent books in English are *Theology and Marxism in Eagleton and Žižek: A Conspiracy of Hope* (2012) and *Heavenly Bodies: Incarnation, the Gaze and Embodiment in Christian Theology* (2016). He has been a research fellow at Uppsala, Cambridge, and Princeton, as well as guest researcher in Nagoya, Stellenbosch, Rome and Oxford.

Fredrik Svenaeus is a professor at the Centre for Studies in Practical Knowledge, Södertörn University, Sweden. His main research areas are philosophy of medicine, bioethics, medical humanities and philosophical anthropology. Current research projects focus on existential questions in association with various medical technologies and on the phenomenology of suffering in medicine and bioethics. He has published widely in these fields; his most recent book is *Phenomenological Bioethics: Medical Technologies, Human Suffering, and the Meaning of Being Alive* published in 2017 by Routledge.

Talia Welsh, PhD, is a U.C. Foundation Professor of Philosophy and Women's Studies at the University of Tennessee Chattanooga. She has published numerous articles, book chapters, a translation of Maurice Merleau-Ponty's lectures in psychology *Child Psychology & Pedagogy: Maurice Merleau-Ponty at the Sorbonne* with Northwestern (2010) and a book *The Child as Natural Phenomenologist: Primal and Primary Experience in Merleau-Ponty's Psychology* also with Northwestern (2013). She is currently finishing a manuscript on the ethics of campaigns to modify the public's health habits.

Index

activity 13–23, 67, 105, 110, 116, 178, 191–2
affect 16, 24, 31–2, 36–9, 53–4, 62, 88, 102, 111–14, 122–3, 124–5
alienation 52–3, 63
ambiguity 35, 44
anxiety 38, 108, 110–11, 193–4, 219–20
assisted reproductive technology/ART 202, 206, 210–12
authenticity 147, 198

Beauvoir, Simone de 209–10
being-in-the-world 63, 109, 164–7, 169–70, 191, 196, 216
Binswanger, Ludwig 29, 32, 42
biological organism 225–8
bio-medical 132, 137, 159, 170, 198
body: body-for-others 196; body schema 37, 52–4, 128, 166–7, 191; dys-appearing body 18, 92–3, 129–30; objective body 2–3, 19, 57; phenomenal body 30–3, 91–3, 216; and suffering 6, 22–4, 75–80, 87–100, 116, 148, 157–70, 182–3, 199–200, 207, 226
brain death 223–8
Butler, Judith 44, 74, 92

Canguilhem, Georges 148, 169
Carel, Havi 17, 20, 166, 169, 177, 179, 181–3
chronic fatigue 113, 140–9
chronic pain 96, 120, 132–3, 142, 147
clinical experience 29–30, 32–3, 36, 39, 43–4
Cole, Jonathan 167
colonialism 50–5, 59–60

colonial violence 49–51
coma 222–3, 227–9
Conrad, Joseph 67–8, 77–80
Conrad, Peter 212

Dasein 15–17, 63, 108–11, 193
death 7–8, 16, 74, 79, 198, 215–29
dehiscence 49, 58–9, 62
dementia 221–4
depression 61, 160
Diagnostic and Statistical Manual of Mental Disorders 145
disease 18, 88, 95–6, 146–7, 174, 176–9, 196, 205, 207, 211–13
dying 174, 194, 215–16, 218–20, 226–7

eidetic 160, 208
embodiment 3–4, 13, 32–3, 43, 58–9, 80, 91–4, 107, 161, 166, 209–10, 215, 221, 224
emotion 36, 61, 88, 106, 119–20, 126, 182, 193–4, 218
enactment 38–40
existence 16–17, 88–91, 98–9, 102, 196, 108–9, 191–4, 200, 226
existential health 87, 96–9

fatigue 6, 103, 113, 137–44, 149
feeling 55, 59, 79–80, 110, 121, 138, 165, 217–18
flesh 58
formal indication 115
Frank, Arthur 177–8
Fuchs, Thomas 31, 158, 160, 162

Gadamer, Hans-Georg 191, 199
Gelb, Adhemar 161
Goldstein, Kurt 125, 161, 165–8

health 164–5, 174, 191, 196, 206–7
healthcare providers 104, 145, 149, 190, 195
health problems 101
heart attack 176, 178, 188–200
Heidegger, Martin 3, 4, 15–20, 63, 108–11, 129, 161, 165, 190, 191, 193–8, 216, 218–21, 226–7
Henry, Michel 4, 23–5, 68
Husserl, Edmund 2–3, 13–16, 19, 23, 130, 160, 163

"I can" 14–15, 19–20, 24, 26, 128, 130, 191–2
"I cannot" 17, 19–20, 26, 130
illness 18, 20, 145–8, 157–9, 164–9, 173–85, 189–91, 196–200, 205–7, 226
imitation 37
infertility 202–3, 206–7, 212–13
injury 55, 119, 124, 129, 166–8, 173–6, 178–83
instrumentalization 8, 53, 67, 73, 75, 78, 199, 224, 229
intentionality 15, 19, 23–4, 27, 33, 52, 56, 88, 92–3, 124–8, 132, 138, 148, 160, 162
interpretation 14, 30–2, 39–43, 67–70, 72–3, 80, 87, 90, 115, 147, 190–1, 193–200
intersubjectivity 6, 35, 42, 53, 58, 61, 69, 72, 75, 77, 79, 146, 157, 158, 163, 195, 203–4, 217
introjection 37–8, 43
in vitro fertilization/IVF 203–10
Irigaray, Luce 203, 209, 212
ISIS 67, 72–4

Jaspers, Karl 158–60

King, Martin Luther 75–8, 80
Klein, Melanie 37–8
Körper 2, 13, 80, 101, 131, 190, 192, 194

Leder, Drew 5, 18, 56, 92–3, 128–30, 149, 163, 173–86
Leib 2, 13, 23, 80, 101, 127, 131, 190–2
Levinas, Emmanuel 4, 68, 70–2, 75, 79–81, 133, 139, 149, 200, 217
life 23–7, 31–2, 68, 80, 101, 104–8, 111, 114–16, 120, 143,

158, 174–6, 180–5, 191–2, 204, 209–10, 213, 215–22, 224–7
lifeworld 2, 16, 77, 105, 107, 159, 164
limitations 5, 102, 110, 130, 168, 174–5, 180–5, 192, 200, 204–5
look (the) 53, 196

markings-out 115
medicalization 145, 202, 212
mental health 95–6, 145, 147, 158–9, 167
Merleau-Ponty, Maurice 2–5, 14, 16–20, 24, 29–44, 49, 51–2, 57–8, 70, 91–2, 119, 121–3, 126–9, 131, 133, 139, 142, 148, 158, 164–7, 169, 178–9, 191, 207, 209–10, 212, 215
mood 106, 108, 110, 165, 193–4

narrative 30, 39–42, 67–9, 71–3, 75, 79–80, 98, 103, 107, 141–2, 197, 218–19, 221–2, 224–9
National Socialism 49–50, 55, 57, 63
negative certainty 116
non-violent resistance 67, 72, 75–6
normal/normality 4–6, 18–20, 22–3, 69, 101, 148, 157–8, 160–70, 176–7, 184, 196, 211–13
not-being-at-home 63, 108–10

object 15, 33–5, 37, 39, 48, 51–4, 57–8, 76, 78, 88–9, 91–2, 101–17, 126–7, 138, 175–6, 178–9, 190–1, 215–16, 228
objectivity 7, 122, 209
Oksala, Johanna 208, 214, 241
organ transplantation 8, 228
orientation 4, 15, 29, 51–2, 56, 59, 64, 128, 159, 170, 232

pain 3–7, 22, 24, 55–6, 58–9, 63, 66, 69, 71, 75, 87–100, 102, 105–8, 112, 115, 119–27, 129–33, 142, 147–9, 188, 197–200
Parnas, Josef 158, 161, 171, 241
Pascal, Blaise 212–14, 241
passion 88–9, 100, 173, 199
passivity 4–7, 13, 15, 16–17, 20–3, 25–6, 30, 41, 45, 47, 60, 69, 77, 87–9, 91, 94–9
pathology 30, 101, 151, 157, 159–65, 169
Paul 6, 13, 20, 21–7

perception 3, 8, 13–14, 24, 29, 33–5, 51, 58, 122–5, 129, 146–7, 207, 211, 224
personhood 8, 80, 97, 216–18, 221–2, 225–8, 230
phenomenology 1–8, 13, 16–17, 20, 26, 29, 37, 43, 58, 63, 67–70, 72, 80, 87, 119, 121, 137, 139, 141, 157–8, 160, 163–5, 170, 189–90, 192, 194–5, 200, 202–4, 215–16
phenomenon 18, 24, 30, 34–5, 37, 42, 63, 97, 101–5, 108–16, 119, 130, 164, 219
positive certainty 103, 116
projection 16–17, 37–9, 193
psychiatry 158–9
psychoanalyses 29, 45
psychopathology 157–61

racism 60
recovery 5, 7, 8, 33, 41–3, 97, 175, 178, 180–1
relatedness 32–3, 35–40, 42–3
responsibility 69, 71–2, 75, 80, 207, 217

Sartre, Jean-Paul 3, 49, 51, 53, 59–61, 140, 148–9, 178, 194, 196
Sass, Louis 158, 160, 171
saturation 2, 114, 117
schizophrenia 158–63
Schneider 18–19, 161–2
second-person perspective 225
selfhood 26, 33, 70–1, 210, 217
shame 38, 53, 218
spatiality 51–2, 92, 129
Steinbock, Anthony 230, 237, 243
stimmung 193
Stoller, Silvia 214, 243
suffering 2, 6–8, 23–4, 44, 62, 75–7, 79–80, 87–91, 93–100, 116, 119, 157, 159, 162–70, 183, 199–200, 221
Svenaeus, Fredrik 5, 8, 164–5, 215, 230–1

technology 179, 195, 200, 202, 224
temporality 8, 138–43, 148, 192, 203, 208
third-person perspective 2, 131, 189, 195, 225
thrownness 109–10
tiredness 7, 102, 113–14, 137–42, 148–9
Toombs, S. Kay 5, 164, 171, 238
torture 6, 49–50, 54–7, 59–63
transference 37
trauma 39–42

uncanny (the) 109, 193, 229
unconsciousness 35, 204

violence 6, 44, 49–51, 54–7, 59–61, 63
vulnerability 1, 5–6, 44, 49, 55, 71–2, 76, 90–1, 99–100, 133, 189, 198, 199–200, 221

weakness 17, 20–7, 195
Winnicott, Donald 36–7, 39–40, 47
world 1–4, 6–7, 14–20, 23–4, 33, 51–2, 54, 57–60, 62–3, 70, 87, 92, 99, 109–11, 121–3, 125–32, 138–9, 157–73, 179, 189, 191, 206–7, 215, 217, 228

Young, Iris Marion 135, 201, 203–4, 209, 213, 244

Zahavi, Dan 161, 171, 216–17, 229

Printed in Great Britain
by Amazon